T0255477

# Lecture Notes in Mathematics 1915

Türker Bıyıkoğlu · Josef Leydold
Peter F. Stadler

# Laplacian Eigenvectors
# of Graphs

## Perron-Frobenius and Faber-Krahn
## Type Theorems

 Springer

Authors

Türker Bıyıkoğlu

Department of Mathematics
Faculty of Arts and Sciences
Işık University
Şile 34980, Istanbul
Turkey
*e-mail: turker.biyikoglu@isikun.edu.tr*
*URL: http://math.isikun.edu.tr/turker*

Josef Leydold

Department of Statistics and Mathematics
Vienna University of Economics
    and Business Administration
Augasse 2-6
1090 Wien
Austria
*e-mail: josef.leydold@wu-wien.ac.at*
*URL: http://statmath.wu-wien.ac.at/~leydold/*

Peter F. Stadler

Bioinformatics Group
Department of Computer Science
University of Leipzig
Härtelstrasse 16-18
04107 Leipzig
Germany
*e-mail: peter.stadler@bioinf.uni-leipzig.de*
*URL: http://www.bioinf.uni-leipzig.de*

Library of Congress Control Number: 2007929852

Mathematics Subject Classification (2000): 05C50, 05C05, 05C35, 05C75, 15A18, 05C22

ISSN print edition: 0075-8434
ISSN electronic edition: 1617-9692
ISBN 978-3-540-73509-0 Springer Berlin Heidelberg New York
DOI 10.1007/978-3-540-73510-6

Springer is a part of Springer Science+Business Media
springer.com
© Springer-Verlag Berlin Heidelberg 2007

Typesetting by the authors and SPi using a Springer LaTeX macro package

Cover design: *design & production* GmbH, Heidelberg

Printed on acid-free paper     SPIN: 12087976     41/SPi     5 4 3 2 1 0

# Preface

Eigenvectors of graph Laplacians are a rather esoteric topic for a book. In fact, we are not aware of even a single review or survey article dedicated to this topic. We have, however, two excuses: (1) There are fascinating subtle differences between the properties of solutions of Schrödinger equations on manifolds on the one hand, and their discrete analogs on graphs. (2) "Geometric" properties of (cost) functions defined on the vertex sets of graphs are of practical interest for heuristic optimization algorithms. Lov Grover's observation that the cost functions of quite a few of the well-studied combinatorial optimization problems are eigenvector of associated graph Laplacians prompted us to investigate such eigenvectors more systematically.

The book in essence covers two topics: Nodal domain theorems which give bounds on the number of connected subgraphs on which an eigenvector does not change sign, and Faber-Krahn-type inequalities which are concerned with the shape of domains (i.e., graphs in our setting) with fixed volume that minimize the first Dirichlet eigenvalue. The connecting theme between these two topics is focus on local and global properties of the eigenvectors (rather than eigenvalues) and convenience of the Rayleigh quotient in the proofs.

The mindful reader will find that more often than not a simple star graph already provides a counterexample for "obvious" conjectures. In fact, we used the Petersen graph just because it seems against tradition to write about graph theory without using the Petersen graph as a counterexample at least once. The simplicity of the counterexamples highlights how little we know about the universe of graph Laplacian eigenvectors (and fitness landscapes in general), and how misguided an intuition trained on well-behaved manifolds can be in this realm: even small moves frequently causes a broken nose caused by some unexpected wall.

The history of this monograph goes back more than a decade and has its roots in the interdisciplinary research environment at the Department of Theoretical Chemistry at the University of Vienna, Austria. A collaboration with Brian Davies during his stay at the Erwin Schrödinger Institute in Vienna in 1995 stimulated our interest in Laplacian eigenvectors and eventually resulted in a research grant from the Austrian *Fonds zur Förderung der Wissenschaftlichen Forschung* (project no. 14094-MAT) to investigate this topic in a more systematic way. Over the years, many colleagues contributed through helpful discussions, among them Wim Hordijk, Jürgen Jost, Bojan Mohar, Tomaž Pisanski, Dan Rockmore, and Gerhard Wöginger. We also thank the Max Planck Institute for Mathematics in the Sciences in Leipzig for their hospitality and for providing a fruitful scientific working for one of us (TB).

Leipzig,                                                          *Türker Bıyıkoğlu*
Wien,                                                              *Josef Leydold*
May 2006                                                       *Peter F. Stadler*

# Contents

# 1

# Introduction

The foundations of spectral graph theory were laid in the fifties and sixties of the 20th century. The eigenvalues of graphs, most often defined as the eigenvalues of the adjacency matrix, have since then received much attention as a means of characterizing classes of graphs and for obtaining bounds on properties such as the diameter, girth, chromatic number, connectivity [14, 17, 45, 46, 83, 85]. The interest has since then shifted somewhat from the adjacency spectrum to the spectrum of the closely related *graph Laplacian* [14, 35, 41, 85, 88, 137, 139]. In particular, Laplacian graph spectra are being investigated as a means of characterizing large "small world networks" and random graphs, see e.g. [33, 34, 119] for a few examples. For the most part, the theory is still concerned with the eigenvalues.

The *eigenvectors* of graphs, on the other hand, have received only sporadic attention on their own, e.g. [134]. Even the book *Eigenspaces of Graphs* by Cvetković et al. [47] contains only a few pages on the geometric properties of the eigenvectors which are mostly used as a convenient proof technique.

In this book we will focus on mostly geometric properties of the eigenvectors themselves. The motivation for this topic is twofold. As we shall see in this first introductory chapter, these objects arise in very diverse applications, from mathematical biology to combinatorial optimization. The Laplacian eigenvectors are used as tools in heuristics to solve combinatorial problems on given graphs, usually without a thorough understanding why they work so well. From a more formal point of view, Laplacian eigenvectors are the natural discretization of eigenfunctions of Laplace-Beltrami operators on manifolds. Surprisingly, some of their properties in the discrete case are reminiscent of corresponding results in the continuous setting, but often there are subtle differences which we found interesting enough to explore in some detail.

We should, at this point, warn the reader: this book collects a number of interesting facets of our topic, enough as we hope to stimulate further research, but it cannot provide a coherent theoretical framework or a powerful machinery to tackle the properties of Laplacian eigenfunctions in generality.

## 1.1 Matrix Representations of a Graph

There are two obvious ways of specifying a simple[1] graph $G(V, E)$ with vertex set $V = \{1, \ldots, n\}$ and edge set $E$ by means of a matrix: the adjacency matrix and the incidence matrix. The *adjacency matrix* $\mathbf{A}$ has entries $A_{xy} = 1$ if $xy \in E$ and $A_{xy} = 0$ otherwise: In order to specify the incidence matrix $\boldsymbol{\nabla}$ we need an arbitrary but fixed orientation (direction) for each edge $e = xy$. Then $\boldsymbol{\nabla}$ is a $(|E| \times |V|)$ matrix and has entries $\nabla_{ex} = -1$ if $x$ is the initial vertex of edge $e$, $\nabla_{ex} = 1$ if $x$ is the terminal vertex of edge $e$, and $\nabla_{ex} = 0$ otherwise, i.e., if $x$ is not in $e$.

Let us now consider a real-valued function $f$ over $V$, $f : V \to \mathbb{R}$. This is simply a vector indexed by the vertices of $G$. In this book we prefer to use a "functional" notation that emphasizes the similarities between the situation of graphs and manifolds. Obviously the set of such functions forms a vector space which is isomorphic to $\mathbb{R}^n$ (and thus we will — by abuse of notation — simply denote this vector space by $\mathbb{R}^n$). Similarly there exists a set of real-valued functions over $E$. The map $f \mapsto \boldsymbol{\nabla} f$ is known as the *co-boundary mapping* of the graph $G$. Its value $(\boldsymbol{\nabla} f)(e)$ at a given edge $e$ is the difference of the values of $f$ at the two end-points of the edge $e$ (considering orientation). Therefore the incidence matrix $\boldsymbol{\nabla}$ is a kind of difference or "discrete differential" operator on $G$.

Let us now consider an Eulerian graph $G$. Recall that $G$ is Eulerian if and only if $G$ is connected and all vertices have even degree. Let $C$ be an (arbitrary) Eulerian cycle in $G$ (i.e., a closed walk that traverses each edge exactly once) and fix an orientation of $G$ such that $C$ is properly oriented in the sense that all edge point "forward" along $C$. The cycle $C$ may pass through each vertex $x$ multiple times; the incoming edge of the $i$-th pass is $e'_i = (y'_i, x)$, the outgoing edge is $e''_i = (x, y''_i)$. We can now define "2nd derivatives" along $C$:

$$(\partial^2_{C;i} f)(x) := (\boldsymbol{\nabla} f)(e'') - (\boldsymbol{\nabla} f)(e')$$
$$= [f(y''_i) - f(x)] - [f(x) - f(y'_i)]$$
$$= f(y'_i) + f(y''_i) - 2f(x) \,.$$

Note that $(\partial^2_{C;i} f)(x)$ is independent of the orientation on $G$. Interpreting each pass of $C$ through $x$ as a different "dimension" it seems natural to consider the sum over these "2nd derivatives" as a "Laplace-Beltrami operator"

$$(\boldsymbol{\Delta} f)(x) = \sum_{\text{passes } i \text{ of } C \text{ through } x} (\partial^2_{C;i} f)(x)$$

$$= \sum_{i=1}^{d(x)/2} [f(y'_i) + f(y''_i) - 2f(x)] = \sum_{y \sim x} [f(y) - f(x)]$$

---

[1] For basic definitions and results the reader is referred to Appendix A.

which is independent of the choice of the Eulerian cycle $C$ and the orientation on $G$. Naturally, one generalizes this definition of $\boldsymbol{\Delta}$ to arbitrary graphs.

In the graph theory literature, however, it is customary to define the Laplacian operator (map) $\mathcal{L} \colon \mathbb{R}^{|V|} \to \mathbb{R}^{|V|}$ with the opposite sign:

$$(\mathcal{L}f)(x) = (-\boldsymbol{\Delta}f)(x) = \sum_{y \sim x}[f(x) - f(y)] \ . \tag{1.1}$$

From an algebraic point of view it appears more natural to define

$$\mathbf{L} = \boldsymbol{\nabla}^{\mathsf{T}}\boldsymbol{\nabla} \tag{1.2}$$

which is known as the *Laplacian (matrix)* of $G$. We have

$$L_{xy} = \sum_{e \in E} \nabla_{ex}\nabla_{ey} = \begin{cases} -1 & \text{if } xy \in E, \\ d(x) & \text{if } x = y, \\ 0 & \text{otherwise,} \end{cases} \tag{1.3}$$

where $d(x) = |\{e \in E | x \in e\}|$ is the *degree* of the vertex $x$. It is important to note that $L_{xy}$ in (1.3) is independent of the orientation of the edges. Clearly, we have the identity $(\mathcal{L}f)(x) = (\mathbf{L}f)(x)$.

Defining the diagonal matrix $\mathbf{D}$ with entries $D_{xx} = d(x)$, called the *degree matrix*, we obtain a simple connection between the Laplacian and the adjacency matrix of a graph,

$$\mathbf{L} = \mathbf{D} - \mathbf{A} \ . \tag{1.4}$$

The Laplacian $\mathbf{L}$ therefore uniquely determines its graph through its off-diagonal entries.

The close relation between $\boldsymbol{\nabla}$ and $\mathbf{L}$ on the one hand, and their differential operator counterparts on the other hand, is exemplified by the following discrete version of *Green's formula*, which is easily verified by direct computation [159]:

**Proposition 1.1.** *Let $f \colon V \to \mathbb{R}$ and $g \colon V \to \mathbb{R}$ be two arbitrary functions. Then*

$$\sum_{x \in V} f(x)\,(\mathbf{L}g)(x) = \sum_{x \in V} g(x)\,(\mathbf{L}f)(x) = \sum_{e \in E}(\boldsymbol{\nabla}f)(e)\,(\boldsymbol{\nabla}g)(e) \ .$$

Using angular brackets $\langle \cdot, \cdot \rangle$ to denote the usual scalar product of two vectors in $\mathbb{R}^n$ and the symbol $-\boldsymbol{\Delta}$ for $\mathbf{L}$ we can formulate Green's formula in a more familiar form as

$$-\langle f, \boldsymbol{\Delta}g \rangle = -\langle \boldsymbol{\Delta}f, g \rangle = \langle \boldsymbol{\nabla}f, \boldsymbol{\nabla}g \rangle \ .$$

The Laplacian $\mathbf{L}$ can thus be viewed as a proper discretization of the usual Laplace-Beltrami differential operator.

## 1.2 Finite Differences

Partial elliptic differential equations play an important rôle in mathematical physics. Examples are the Poisson equation

$$\Delta u = f \quad \text{on } \Omega$$

with given domain $\Omega \subseteq \mathbb{R}^s$ and $f \in \mathcal{C}^0(\Omega)$, the eigenvalue problem

$$-\Delta u = \lambda u \quad \text{on } \Omega$$

or Schrödinger's equation. Here $\Delta$ denotes the classical Laplace operator given as $\Delta u = \sum_{i=1}^{s} u_{x_i x_i}$.

Computing solutions of such differential equations is a challenging task in numerical mathematics. An old and popular method is based on *finite differences*: A grid or mesh is used to divide $\mathbb{R}^s$ into small hyper-rectangles or simplices. At the nodes of the grid the Laplace operator $\Delta$ is approximated by a difference operator. For $\Omega \subseteq \mathbb{R}^2$ and a square mesh of width $h$ we get the so called *Five-Point Formula*

$$\Delta_h u(x,y) = [u(x+h,y) + u(x,y+h) + u(x-h,y) + u(x,y-h) - 4\,u(x,y)]/h^2 \,.$$

From a graph theoretical point of view the square mesh is the Cartesian product of two paths of proper lengths, $P_{k_1} \Box P_{k_2} \subset \mathbb{Z}^2$, and $\Delta_h$ is the graph Laplacian $\mathbf{L}(P_{k_1} \Box P_{k_2})$ times the constant $-1/h^2$. Thus the graph Laplacian arises in a quite natural way. For details about finite differences (and other methods for solving elliptic partial differential equations) the interested reader is referred to [92], or to [120] for the special case of Laplacian eigenvalues.

## 1.3 Landscapes on Graphs

Maybe the most direct interest in the structure of the *eigenfunctions* of graph Laplacians comes from the theory of *fitness landscapes*, see [149] for a review. Evolution theory has as its cornerstone the concept of *fitness*. Fitness is traditionally defined as the relative reproductive success of a genotype as measured by survival, fecundity or other life history parameters [27, 96, 165]. The key principle of Darwinian evolutionary theory is that natural selection acts so as to (locally) maximize the fitness of a species or population. The concept of a *fitness landscape* originated in the 1930s in theoretical biology [177, 178] as a means of visualizing this kind evolutionary adaptation: A fitness landscape is a kind of "potential function" on which a population moves uphill due to the combined effects of mutation and selection. Thus, natural selection can be viewed as a type of "hill climbing" on the topography implied by the fitness function.

Models of disordered systems, in particular spin glasses, naturally led to the notion of landscapes [18, 135]: Each spin configuration is assigned an energy by virtue of the Hamiltonian that specifies the model. In the simplest case so-called Ising spins are considered, which can only take two values: *up* ($\sigma = +1$) and *down* ($\sigma = -1$). The Hamiltonian of the system typically considers the interactions between neighboring spins, in the simplest case

$$f(\boldsymbol{\sigma}) = \sum_{\text{neighbors } i,\, j} \sigma_i \sigma_j \ .$$

There is also a close conceptual similarity of the landscapes in biology and spin glass physics with the *potential energy surfaces* (PES) of theoretical chemistry [95, 136].

In combinatorial optimization the fitness function $f$ is usually referred to as the *cost function* on a *search space* $X$ [80]. The Traveling Salesman Problem (TSP) is probably the most frequently studied combinatorial optimization problem. The ingredients of the TSP are simple enough: The configurations are the $n!$ permutations of the $n$ locations, usually called a "tour". We write $\pi = (\pi(1), \ldots, \pi(n))$ for the order in which they are visited. Given the travel distance (or cost) $C_{kl}$ from city $l$ to city $k$ we can write down the cost function in the form

$$f(\pi) = \sum_{i=1}^{n-1} C_{\pi(i+1),\pi(i)} + C_{\pi(1),\pi(n)} \ ,$$

where the last term describes returning to the point of origin.

In formal terms, a *landscape* is a triple $(X, \mathcal{X}, f)$ consisting of:

1. A set $X$ of configurations,
2. a notion $\mathcal{X}$ of neighborhood, nearness, distance, or accessibility on $X$, and
3. a fitness function $f \colon X \to \mathbb{R}$.

The set $X$ together with the "structure" $\mathcal{X}$ forms the configuration space. In the simplest case, $\mathcal{X}$ describes which configurations can be obtained from a given one by means of basic "moves" or transformations. Examples of such moves are the flipping of a single spin, the exchange of a single letter by another one in a genetic sequence, or the transposition of two cities along the salesman's tour. Usually the move-set is constructed in a symmetric way, so that the configuration space $(X, \mathcal{X})$ becomes an undirected finite graph $G$. More general classes of configuration spaces are discussed e.g. in [156].

Let us consider the function $\tilde{f}$ given by $\tilde{f}(x) = f(x) - \bar{f}$, where $\bar{f} = \frac{1}{|X|} \sum_{x \in X} f(x)$ is the average cost of an arbitrary configuration. Grover and others [37, 90, 157] observed that $\tilde{f}$ is in many cases an eigenfunction of the Laplacian $\mathbf{L}$ of the graph representing the configuration space $(X, \mathcal{X})$. These landscapes have been termed *elementary* in [157]. Some examples are collected in Table 1.1.

Lov Grover [90] showed that, if $f$ is an elementary landscape, then

$$f(\hat{x}_{\min}) \le \bar{f} \le f(\hat{x}_{\max})$$

**Table 1.1.** Examples of Elementary Landscapes

| Problem | Graph | degree | $\lambda$ | Order | Reference |
|---|---|---|---|---|---|
| $p$-spin glass | $\mathcal{Q}_2^n$ | $n$ | $2p$ | $p$ | definition |
| NAES | $\mathcal{Q}_2^n$ | $n$ | $4$ | $2$ | [90] |
| Weight Partitioning | $\mathcal{Q}_2^n$ | $n$ | $4$ | $2$ | [90, 157] |
| GBP (constrained) | $\mathcal{Q}_2^n$ | $n$ | $4$ | $2$ | [2] |
| Max Cut | $\mathcal{Q}_2^n$ | $n$ | $4$ | $2$ | [2] |
| Graph $\alpha$-Coloring | $\mathcal{Q}_\alpha^n$ | $(\alpha-1)n$ | $2\alpha$ | $2$ | [157] |
| XY-spin glass | $\mathcal{Q}_\alpha^n$ | $(\alpha-1)n$ | $2\alpha$ | $2$ | [79] |
| for $\alpha > 2$: | $\mathcal{C}_\alpha^n$ | $2$ | $8\sin^2(\pi/\alpha)$ | $2$ | [79] |
| Linear Assignment | $\Gamma(\mathsf{S}_n, \mathcal{T})$ | $n$ | | $1$ | [151] |
| TSP symmetric | $\Gamma(\mathsf{S}_n, \mathcal{T})$ | $n(n-1)/2$ | $2(n-1)$ | $2$ | [37, 90] |
| | $\Gamma(\mathsf{S}_n, \mathcal{J})$ | $n(n-1)/2$ | $n$ | $2$ | [37, 90] |
| antisymmetric | $\Gamma(\mathsf{S}_n, \mathcal{T})$ | $n(n-1)/2$ | $2n$ | $3$ | [11, 157] |
| | $\Gamma(\mathsf{S}_n, \mathcal{J})$ | $n(n-1)/2$ | $n(n+1)/2$ | $\mathcal{O}(n)$ | [11, 157] |
| Graph Matching | $\Gamma(\mathsf{S}_n, \mathcal{T})$ | $n(n-1)/2$ | $2(n-1)$ | $2$ | [157] |
| Graph Bipartitioning | $J(n, n/2)$ | $n^2/4$ | $2(n-1)$ | $2$ | [90, 161, 162] |

Here $\mathcal{Q}_\alpha^n$ is a Hamming graphs, i.e., the $n$-fold Cartesian product of the complete graph $K_\alpha$, $\Gamma(\mathsf{A}, \Omega)$ is the Cayley graph of the group $\mathsf{A}$ with generating set $\Omega$, where $\mathsf{S}_n$ and $\mathsf{A}_n$ denotes the symmetric and alternating groups, resp., $\mathcal{T}$, $\mathcal{J}$, and $\mathcal{C}_3$ are the transpositions, reversals, and permutations defined by a cycle of length 3, resp. $J(p,q)$ is a Johnson graph. The order of eigenvalue $\lambda$ is its position in the spectrum of $\mathbf{L}$ without counting multiplicities and defining the order of $\lambda = 0$ as 0.

where $\hat{x}_{\min}$ and $\hat{x}_{\max}$ are arbitrary local minima and maxima, respectively. This *maximum principle* shows that elementary landscapes are well-behaved: There are no local optima with worse than average fitness $\bar{f}$.

Many of the examples in Table 1.1 belong to the first few eigenvalues of $\mathbf{L}$. A simple relationship between $\lambda$ and the autocorrelation function of $f$ of $(X, \mathcal{X})$, see e.g. [157], suggests furthermore that the "ruggedness" [72, 173] of an elementary landscape, and hence its difficulty for evolutionary adaptation, should be related to its corresponding eigenvalue $\lambda$ of $\mathbf{L}$. Furthermore, a Fourier-decomposition-like formalism was developed that decomposes arbitrary landscapes into their elementary components [98, 151, 174]:

$$f = a_0 + \sum_{k>0}^{n-1} a_k f_k \tag{1.5}$$

where the $f_k$ form an orthonormal system of eigenfunction of the graph Laplacian, $\mathbf{L}f_k = \lambda_k f_k$, and $a_0 = \bar{f}$ is the average value of the function $f$. Let us denote the distinct eigenvalues of $\mathbf{L}$ by $\bar{\lambda}_p$, sorted in increasing order starting with $\bar{\lambda}_0 = \lambda_0 = 0$. We call $p$ the *order* of the eigenvalue $\bar{\lambda}_p$. The *amplitude spectrum* of $f: V \to \mathbb{R}$ is defined by

$$B_p = \sum_{k:\, \lambda_k = \bar{\lambda}_p} |a_k|^2 \Big/ \sum_{k>0} |a_k|^2 . \tag{1.6}$$

By definition, $B_p \geq 0$ and $\sum_p B_p = 1$. The amplitudes measures the relative contribution of the eigenspace of the eigenvalue with order $p$ to the function $f$. Of course, a landscape is elementary if and only if $B_p = 1$ for a single order and 0 for all others.

## 1.4 Related Matrices

Let us briefly mention a few matrices that are closely related to $\mathbf{L}$. Sometimes a normalized version $\mathbf{L}^*$ representing the average difference between $x$ and its neighbors is used:

$$(\mathbf{L}^* f)(x) = \frac{1}{d(x)} \sum_{y \sim x} [f(x) - f(y)] .$$

This definition is quite similar to (1.1). In fact for graphs without isolated vertices we have

$$\mathbf{L}^* = \mathbf{D}^{-1}\mathbf{L} = \mathbf{I} - \mathbf{D}^{-1}\mathbf{A} .$$

This version is used e.g. by Grover [90] and Barnes and coworkers [12, 155]. The first nontrivial eigenvalue of $\mathbf{L}^*$ plays an important role for synchronization in coupled map lattices [5, 105].

Chung [35] defined a general and normalized form of the Laplacian matrix, which is consistent with the eigenvalues in spectral geometry and in stochastic processes:

$$\tilde{L}_{xy} = \begin{cases} 1 & \text{if } x = y \text{ and } d(x) > 0, \\ -1/\sqrt{d(x)d(y)} & \text{if } xy \in E, \\ 0 & \text{otherwise.} \end{cases}$$

In matrix form we have $\tilde{\mathbf{L}} = \mathbf{D}^{-1/2}\mathbf{L}\mathbf{D}^{-1/2}$ for graphs without isolated vertices. $\tilde{\mathbf{L}}$ and $\mathbf{L}^*$ are similar for graphs without isolated vertices: $\mathbf{L}^* = \mathbf{D}^{-1/2}\tilde{\mathbf{L}}\mathbf{D}^{1/2}$.

Another associated matrix is the transition operator $\mathbf{T} = \mathbf{A}\mathbf{D}^{-1}$ of an unbiased random walk on $G^2$. We have therefore

$$\mathbf{L}^* = \mathbf{I} - \mathbf{T}^{\mathsf{T}} \qquad \text{and hence} \qquad (\mathbf{L}^*)^{\mathsf{T}} = \mathbf{I} - \mathbf{T}$$

as the associated "Laplacian". This version is used e.g. in [164]. The matrices $\mathbf{L}^*$ and $\mathbf{T}$ are – in contrast to Chung's Laplacian $\tilde{\mathbf{L}}$ – not symmetric unless

---

[2] Contrary to the convention in the Markov chain literature we treat the distributions as column vectors here, i.e., a step of the Markov chains reads $p' = \mathbf{T}p$.

the graph $G$ is regular; hence they do not belong to the class of operators that we will be concerned with in this book.

As an example for the application of the transition operator we briefly continue our discussion of fitness landscapes of the previous section. Let $G$ be a $D$-regular graph and let $f: V \rightarrow \mathbb{R}$ be an arbitrary fitness function. Weinberger [173] suggested to characterize a fitness landscape by means of the autocorrelation function $r$ of the values $f(x)$ sampled along a random walk on $G$. One easily verifies the following relation between $r(s)$ and the Laplacian spectrum [157]:

$$
\begin{aligned}
r(s) &= \frac{E[f(x_{t+s})f(x_t)] - E[f(x_{t+s})]E[f(x_t)]}{E[f(x_t)^2] - E[f(x_t)]^2} \\
&= \langle \tilde{f}, \mathbf{T}^s \tilde{f} \rangle / \langle \tilde{f}, \tilde{f} \rangle \\
&= \sum_{p>0} B_p \left(1 - \lambda_p/D\right)^s
\end{aligned}
\tag{1.7}
$$

Here, the expectation $E[.]$ is taken over all random walks with transition matrix $\mathbf{T}$, all times $t$, and all initial conditions $x_0 \in V$. The autocorrelation function $r(s)$ is therefore a superposition of exponential functions. It decays more rapidly, when the amplitudes $B_p$ with large Laplacian eigenvalues $\lambda_p$ increase. A landscape is therefore elementary if and only if its autocorrelation function decays exponentially. The correlation length

$$
\ell := \sum_{s=0}^{\infty} r(s) = D \sum_{p>0} B_p/\lambda_p
\tag{1.8}
$$

also reflects the fact that the "smoothness" or ruggedness of a fitness landscape is directly related to the amplitude spectrum. The correlation length of an elementary landscape is therefore determined by the order $p$ of the associated Laplacian eigenvalue.

## 1.5 Graphs with a Boundary: The Discrete Dirichlet Problem

In 1966 Kac [106] asked whether it is possible to *hear the shape of a drum*. A mathematical drum is a domain $D$ with a boundary $\partial D$ in some $\mathbb{R}^n$ (or more generally in some manifold $\mathcal{M}$). If small vibrations are induced in the membrane, it is not unreasonable to expect a point on its surface to move only vertically. In the absence of damping the motion of the point is given by the wave equation

$$
\Delta u + \lambda u = 0
$$

with the constraint that $u(x) = 0$ for all $x \in \partial D$ (the so-called *Dirichlet boundary condition*). Here $\Delta$ denotes the Laplace-Beltrami operator. The solution of a Dirichlet problem involves a countable sequence of eigenvalues (in this case the frequencies of the tones produced by the membrane). Kac's question thus can be rephrased in a more formal way: Can nonisometric drums $D$ afford the same set of eigenvalues? The answer was given in 1992: We cannot hear the shape of a drum, i.e., there are nonisometric domains $D$ that yield the same spectrum [86].

Fisher [70] considered the discrete analog to Kac's problem. In his model the membrane consists of a set of atoms which in the equilibrium state lie on the vertices of a regular lattice graph embedded in a plane. Each atom acts on its neighboring atoms by elastic forces. The discretization of the vibration of a membrane is the Laplacian matrix $\mathbf{L}$ of the graph $G$. The eigenvalues of $\mathbf{L}$ again correspond to the frequencies of the membrane. We also can't hear the discrete shape of a drum, because the eigenvalues of a graph do not determine the graph uniquely; see e.g. [45]. Nevertheless, in practice it is often possible to obtain at least good approximations of a graph (in terms of the cardinality of the symmetric difference between the true graph and its reconstruction) from its spectrum [103].

We need a notion of a graph with boundary for defining discrete analogs of Dirichlet boundary conditions. Of course, graphs do not have boundaries by themselves. Starting from a graph $G(V, E)$ we may, however, consider the induced subgraph $G[V^\circ]$ on a subset $V^\circ \subset V$, considering $V \setminus V^\circ$ as the boundary of $G[V^\circ]$ on which the constraint $u(x) = 0$ is enforced. We denote this boundary by $\partial V$. Formally we can define a *graph with boundary* as a graph $G(V^\circ \cup \partial V, E^\circ \cup \partial E)$ where $V^\circ$ denotes the set of *interior* vertices and $\partial V$ the set of *boundary* vertices. The set of edges between interior vertices are called *interior* edges and denoted by $E^\circ$; edges between $V^\circ$ and $\partial V$ are called *boundary* edges and denoted by $\partial E$. Edges between boundary vertices do not make sense in our setting and are thus deleted. It must be noted here that a graph with boundary is called connected if the graph induced by its interior vertices, $G[V^\circ]$, is connected. The partition into interior and boundary vertices might be to some extend "arbitrary". In the case of drums, however, we might also choose to use a nail and fix the position of the membrane at an arbitrary point, thereby adding an additional point to the boundary of the domain of the corresponding Dirichlet problem. A more thorough discussion of Dirichlet problems on graphs will be given in Chap. 6.

An interesting application of the first Dirichlet eigenvalue arises from a combinatorial game called *chip firing* [23]: Every vertex of a connected graph contains an integral number of chips. In each step of the game a vertex is selected that has at least as many chips as its degree and one chip is moved to each of its neighbors. The game can continue as long as there is a vertex with sufficiently many chips on it. The game terminates when no vertex can be selected. Chung and Ellis [32] considered a variant of this game, in which chips are removed from the game when they are moved across a boundary and gave an upper bound that depends on the first Dirichlet eigenvalue for the number of steps until such a game terminates.

If we restrict ourselves to solutions $f$ of the Dirichlet problem on a graph $G(V^\circ \cup \partial V, E^\circ \cup \partial E)$ with boundary we have to look for a function $f$ which vanishes on all boundary vertices, i.e. $f(x) = 0$ for $x \in \partial V$, and which satisfies for all interior vertices $x \in V^\circ$

$$(\mathbf{L}f)(x) = \sum_{y \sim x} [f(x) - f(y)] = \sum_{y \in V} L_{xy} f(y) = \sum_{y \in V^\circ} L_{xy} f(y) = \lambda f(x)$$

for some eigenvalue $\lambda$. Thus the Dirichlet problem can be reduced to a matrix eigenspace problem for $G[V^\circ]$. The corresponding *Dirichlet matrix* $\mathbf{L}^\circ(G)$ can be derived from the graph Laplacian $\mathbf{L}(G)$ simply by deleting all rows and columns that correspond to boundary vertices, i.e., by using the principal submatrix corresponding to interior vertices. Compared to the "free" graph Laplacian $\mathbf{L}(G[V^\circ])$ on the graph induced by its interior vertices, $G[V^\circ]$, the Dirichlet matrix differs just by an additional "potential" $p(x)$ in the diagonal elements:

$$\mathbf{L}^\circ(G) = \mathbf{L}(G[V^\circ]) + \mathbf{P} \qquad (1.9)$$

where $\mathbf{P}$ is a diagonal matrix whose entries are $P_{xx} = p(x) = |\{y : yx \in \partial E\}|$. For a more "natural" motivation of this definition we refer the interested reader to [75] or Sect. 2.4.

## 1.6 Generalized Graph Laplacians

The Dirichlet operator and Chung's Normalized Laplacian motivate the definition of a more general class of matrices associated with a graph $G(V, E)$. We call a symmetric matrix $\mathbf{M}$ a *generalized Laplacian* or *discrete Schrödinger operator* of $G$ if $M_{xy} < 0$ whenever $xy$ is an edge of $G$ and $M_{xy} = 0$ whenever $x$ and $y$ are distinct and not adjacent. There are no constraints on the diagonal entries of $\mathbf{M}$. Fiedler [67] and Roth [152] call such matrices *"essentially non-positive"*. The ordinary Laplacian $\mathbf{L}$ as well as the negative adjacency matrix $-\mathbf{A}$ are of course generalized Laplacians.

Such generalized graph Laplacians can be interpreted in two ways. First the off-diagonal entries can be seen as coefficient of a discrete analog of an elliptic operator which are used in mathematical physics to describe oscillation in nonhomogeneous matter. On the other hand it could be seen as "ordinary" Laplacian on a weighted graph. Then the weights $w_{xy}$ on an edge $xy$ has to taken into consideration. Thus we have a Hamiltonian operator $\mathcal{H}$ of the form

$$(\mathcal{H}f)(x) = \sum_{y \sim x} w_{xy} [f(x) - f(y)] + p(x) f(x) .$$

The first part of the right hand side then represents the kinetic part while $p(x)$ represents some potential. This is the analogous expression to (1.1) of some generalized Laplacian.

Another graph Laplacian comes from quantum chemistry. The properties of micro-objects (electrons, atoms, molecules) in a stationary state are described by wave functions $\Psi$, representing solutions of Schrödinger's equation $\hat{H}\Psi = E\Psi$, in which $\hat{H}$ is the energy operator and $E$ is the energy of the object under consideration. For a certain class of organic compounds, which includes the conjugated hydrocarbons, one can approximate Schrödinger's equation in the following form

$$\mathbf{H}\,\psi = E\psi$$

where the off-diagonal entries of the matrix $\mathbf{H}$ are the so-called "resonance integrals" describing the interactions between orbitals at different atoms, while the diagonal terms incorporate the so-called "Coulomb integrals". The off-diagonal elements vanish unless the corresponding atoms are connected by bonds, i.e., adjacent in the graph representation of the molecule. The entries of $\mathbf{H}$ are tabulated for different atom and bond types. These approximations form the basis of the so-called *Hückel theory* [101]. The matrix $\mathbf{H}$ is therefore called the *Hückel matrix* of the molecule. Properties of the molecule can now be derived from the eigenvalues and eigenfunctions of $\mathbf{H}$. In particular, the electron density within a given molecular orbital at an atom $x$ is given by $|\psi(x)|^2$. Recently, a Hückel-theory like approach was used to implement a very general artificial chemistry for the purpose of simulating large-scale chemical networks [16]. We refer to [6, 24, 168] for comprehensive presentations of chemical graph theory.

## 1.7 Colin de Verdière Matrices

A special class of generalized graph Laplacians has been of particular interest because of its close relationship with embeddability properties of the associated graphs. While not directly related to the structure of the Laplacian eigenfunctions, we briefly mention this topic for its importance in algebraic graph theory.

For a given graph $G$ consider the class of matrices with the following properties, see e.g. [97]:

1. $\mathbf{M}$ is a generalized Laplacian of $G$.
2. The only symmetric matrix $\mathbf{X}$ with entries $X_{ij} = 0$ whenever $i = j$ or $i$ and $j$ are adjacent in $G$ that satisfies $\mathbf{MX} = \mathbf{O}$ is the zero-matrix $\mathbf{X} = \mathbf{O}$. This property is known as the *Strong Arnold Property*.
3. $\mathbf{M}$ has exactly one negative eigenvalue, which is simple.

Yves Colin de Verdière [38, 40] introduced the parameter $\mu(G)$ as the maximal dimension of the null-space of a matrix satisfying these three properties. This parameter, which is minor-monotone, determines embeddability properties of the graph $G$ [40, 129, 170]:

$\mu(G) \leq 1 \Leftrightarrow G$ is disjoint union of paths
$\mu(G) \leq 2 \Leftrightarrow G$ is outerplanar
$\mu(G) \leq 3 \Leftrightarrow G$ is planar
$\mu(G) \leq 4 \Leftrightarrow G$ is linklessly embeddable

Related results and similar, embedding-related, graph parameters are discussed e.g. in [10, 130, 169].

## 1.8 Practical Applications of Laplacians Eigenvectors

There are many applications and results on graph Laplacian eigen*values* and their relations to numerous graph invariants, including connectivity, expanding properties, genus, diameter, mean distance, and chromatic number, as well as to partition problems (graph bisection, connectivity and separation, isoperimetric numbers, maximum cut, clustering, graph partition), and approximations for optimization problems on graphs (cutwidth, bandwidth, min-$p$-sum problems, ranking, scaling, quadratic assignment problem). For an overview of such results we refer to Mohar's surveys, e.g. [138], or the monographs by Colin de Verdière [41] and Chung [35].

So far we have given examples where eigen*functions* arise in a natural albeit sometimes surprising manner. In this section we collect a few practical applications of eigenfunctions where the appearance of a graph Laplacians is not obvious.

Eigenfunctions corresponding to the second smallest eigenvalue of a Laplacian can be used for graph bipartitioning. In one variant of graph bipartitioning one attempts to find a vertex separator $S$ of the given graph $G$ such that $S$ has few vertices and disconnects $G - S$ into two parts $A$ and $B$ with nearly equal numbers of vertices. Pothen et al. [146] give the following heuristic method for the bipartition: Compute the eigenfunction $f_2$ of the second smallest eigenvalue of the Laplacian. Assign each vertex the value of $f_2$. Compute the median of all these values. Bipartition the vertices as follows: the vertices whose values are less than or equal to the median form one part; the rest of the vertices form the other part. The quality of this heuristic is investigated by Guattery and Miller [91]. Alpert et al. [1] describe a multiple eigenfunction extension of this approach.

Eigenfunctions can also be used to obtain colorings of a graph. For a given collection $\mathcal{F}$ of eigenfunctions, the vertices $u$ and $v$ belong to the same color class (partition) if and only if $u$ and $v$ have the same sign for each $f_k \in \mathcal{F}$. Aspvall and Gilbert [4] describe a procedure in which, starting from the eigenfunctions to the largest eigenvalues of $\mathbf{L}$, additional eigenfunctions are added to $\mathcal{F}$ until one obtains a valid coloring.

A related application is "spectral clustering" [15, 57], which is based on the observation that nodal domains (see Sect. 3.1) of the first eigenvectors of the graph Laplacian can be used as indicators for suitably size-balanced minimum cuts, see e.g. [154]: Consider an undirected edge-weighted graph with weight function $w(x, y) > 0$ if $\{x, y\}$ is an edge in $G$. The weight of a cut $A, X \setminus A$ and the within-group association in $A$ are defined as

$$c(A) = \sum_{x \in A, y \in X \setminus A} w(x, y) \quad \text{and} \quad a(A) = \sum_{x \in A, y \in V} w(x, y),$$

respectively and set $d(x) = \sum_{y \in X} w(x, y)$. A convenient normalized cut-weight is $\nu(A) = c(A)/a(A) + c(A)/a(X \setminus A)$. Instead of using the vertex sets $A$ and $X \setminus A$ we can use a sign-vector $f$ to describe the cut by setting $f(x) = 1$ if $x \in A$ and $f(x) = -1$ if $x \in X \setminus A$. Then a short computation in [154] shows that

$$\min_f \nu(f) = \min_g \frac{\langle g, (\mathbf{D} - \mathbf{W})g \rangle}{\langle g, \mathbf{D}g \rangle}$$

with the constraint $g\mathbf{D}\mathbf{1} = 0$. It follows that the eigenvector of the 2nd-smallest eigenvalue of the symmetric matrix $\tilde{\mathbf{L}} = \mathbf{D}^{-1/2}(\mathbf{D} - \mathbf{W})\mathbf{D}^{-1/2}$ is the appropriate real-valued approximation to the discrete optimization problem for the sign-vector $f$. In practice, one often proceeds by computing the nodal domains of $g$ and then re-iterating the procedure for the corresponding subgraphs.

A different field of applications for spectral methods is graph drawing. An embedding of a graph in $\mathbb{R}^k$ is a map from vertices of a graph into $\mathbb{R}^k$, in other words an embedding consists of the positions of the vertices in a $k$-dimensional drawing of a graph. Eigenfunctions of the second, third, and fourth (sometimes fifth) smallest eigenvalues of a Laplacian can be used for an embedding of a graph in $\mathbb{R}^3$. For a better embedding of fullerenes, the eigenfunctions of the adjacency matrix are also used (see [74, 87, 121, 122]). Pisanski and Shawe-Taylor [144] give a method for drawing a graph in any number of dimensions: Compute an orthonormal basis of eigenfunctions $f_1, \ldots, f_n$ of the Laplacian matrix of a graph with $n$ vertices. The eigenfunctions $f_2, \ldots, f_{k+1}$ yields the columns of the embedding in $\mathbb{R}^k$ with minimum energy (energy is defined as the sum of the square of the Euclidean distance of two adjacent vertices). Pisanski and Žitnik [145] in their short survey on graph representations call this method the *Laplace method* and remark that there is a connection between the number of nodal domains of the eigenfunctions (see Chap. 3) and the quality of the corresponding graphical representation. In bioinformatics, the program SpectralNet [73], which is based on these principles, is used for analyzing and visualizing these biological and chemical networks. Related work is reported by Koren [117]. Lovász and Schrijver [130] gave some results on embeddings of path, 2-connected outerplanar graphs and 3-connected planar graphs based on Colin de Verdière's [38] new graph invariant $\mu$ which is related to generalized graph Laplacians. In their proof they used results related to the nodal domain theorem (Thm. 3.1).

Another application of graph eigenfunctions is used in economics. Maslov [131] gives a simple measure of the level of financial globalization of a given country based on the analysis of cross-correlations between stock market indices in different countries and regions of the world. He studies the empirical correlation matrix (this matrix is symmetric, nonnegative definite but in general it is not a graph Laplacian) of index price fluctuations in a large number of individual countries. The three largest eigenvalues and the corresponding eigenfunctions are used for the observation of the influence of world index dynamics.

# 2

# Graph Laplacians

In this chapter we recall the definition of (generalized) graph Laplacians and present the basic properties of their eigenfunctions. Moreover, we establish the main tools that will be used throughout the book. For a thorough overview of other properties of graph Laplacians not required for our investigations of eigenfunctions we refer the interested reader to the survey by Merris [133].

## 2.1 Basic Properties of Graph Laplacians

Let $G(V, E)$ be a simple graph with vertex set $V$ and edge set $E$. We use the convention that $|V| = n$ and $|E| = m$, i.e., $G$ is a graph with $n$ vertices and $m$ edges. The *Laplacian* of $G$ is the matrix

$$\mathbf{L}(G) = \mathbf{D}(G) - \mathbf{A}(G) \tag{2.1}$$

where $\mathbf{D}(G)$ is the diagonal matrix whose entries are the degrees of the vertices of $G$, i.e. $D_{vv} = d(v)$, and $\mathbf{A}(G)$ denotes the adjacency matrix of $G$. For the function $\mathbf{L}f$ we find

$$(\mathbf{L}f)(x) = \sum_{y \sim x} [f(x) - f(y)] = d(x) f(x) - \sum_{y \sim x} f(y) . \tag{2.2}$$

We denote the eigenvalues of $\mathbf{L}$ by $\lambda_i$ enumerated in increasing order, i.e.,

$$0 = \lambda_1 \leq \lambda_2 \leq \cdots \leq \lambda_n . \tag{2.3}$$

The quadratic form of the graph Laplacian can be computed via Green's formula as

$$\langle f, \mathbf{L}f \rangle = \sum_{x,y \in V} L_{xy} f(x) f(y) = \sum_{xy \in E} (f(x) - f(y))^2 . \tag{2.4}$$

This equality immediately shows that the graph Laplacian is a nonnegative operator, i.e., all eigenvalues are greater than or equal to 0.

A symmetric matrix $\mathbf{M}(G)$ is called a *generalized Laplacian* (or *discrete Schrödinger operator*) of $G$ if it has nonpositive off-diagonal entries and for $x \neq y$, $M_{xy} < 0$ if and only if the vertices $x$ and $y$ are adjacent. On the other hand, for each symmetric matrix with nonpositive off-diagonal entries there exists a graph where two distinct vertices $x$ and $y$ are adjacent if and only if $M_{xy} < 0$. Similarly to (2.2) we have

$$(\mathbf{M}f)(x) = \sum_{y \sim x} (-M_{xy})[f(x) - f(y)] + p(x)\, f(x) \,, \tag{2.5}$$

where $p(x) = M_{xx} + \sum_{y \sim x} M_{xy}$. The last part $p(x)$ can be viewed as some potential on vertex $x$. Defining a matrix $\mathbf{W}$ consisting of $W_{xy} = M_{xy}$ for $x \neq y$ and $W_{xx} = -\sum_{y \neq x} M_{xy}$ and a diagonal matrix $\mathbf{P}$ with the potentials $p(x)$ as its entries we can decompose every generalized Laplacian as

$$\mathbf{M} = \mathbf{W} + \mathbf{P} \,.$$

$\mathbf{W}$ can be seen as *discrete elliptic operator*. The quadratic form of the generalized Laplacian can then be computed as

$$\langle f, \mathbf{M}f \rangle = \sum_{xy \in E} (-M_{xy})(f(x) - f(y))^2 + \sum_{x \in V} p(x)\, f(x)^2 \,; \tag{2.6}$$

an alternative presentation is

$$\langle f, \mathbf{M}f \rangle = \sum_{x \in V} M_{xx} f(x)^2 + 2 \sum_{xy \in E} M_{xy} f(x) f(y) \,. \tag{2.7}$$

The following remarkable result for the eigenvalues of a generalized Laplacian can be easily derived.

**Theorem 2.1 ([22]).** *Let $\lambda$ be an eigenvalue of a generalized Laplacian $\mathbf{M} = \mathbf{W} + \mathbf{P}$ with eigenfunction $f$. Then either $\sum_{v \in V} f(v) = \sum_{v \in V} p(v)\, f(v) = 0$, or*

$$\lambda = \frac{\sum_{v \in V} p(v)\, f(v)}{\sum_{v \in V} f(v)} \,.$$

*Proof.* Let $\mathbf{1} = (1, \ldots, 1)^{\mathsf{T}}$. Then a straightforward computation gives

$$
\begin{aligned}
\langle \mathbf{1}, \mathbf{M}f \rangle &= \sum_{v \in V} \left( \sum_{w \sim v} (-M_{vw})(f(v) - f(w)) + p(v)\, f(v) \right) \\
&= \sum_{v,w \in V} (-M_{vw})(f(v) - f(w)) + \sum_{v \in V} p(v)\, f(v) \\
&= \sum_{v,w \in V} M_{vw} f(w) - \sum_{v,w \in V} M_{vw} f(v) + \sum_{v \in V} p(v)\, f(v) \\
&= \sum_{v \in V} p(v)\, f(v) \,.
\end{aligned}
$$

Since $f$ is an eigenfunction we find $\langle \mathbf{1}, \mathbf{M}f \rangle = \lambda \sum_{v \in V} f(v)$, and thus the proposition follows. $\qquad \square$

*Remark 2.2.* The case $\sum_{v \in V} f(v) = 0$ happens, for example, for all eigenfunctions corresponding to an eigenvalue $\lambda > \lambda_1$ when the eigenfunction $f_1$ of $\lambda_1$ is constant. This is the case if and and only if $p(v)$ is constant for all $v \in V$.

The spectrum of the (generalized) Laplacian provides quite detailed information on the structure of the underlying graph. We refer the interested reader to classical books and surveys, e.g. [17, 35, 41, 46, 85, 133, 137].

One of these basic results is related to the multiplicity of the first eigenvalue and the connectivity of the graph. Notice, that all eigenvalues of a discrete elliptic matrix $\mathbf{W}$ are nonnegative as an immediate consequence of (2.6). Moreover, its smallest eigenvalue is $\lambda_1 = 0$.

**Theorem 2.3.** *Let $\mathbf{W}(G)$ be a generalized Laplacian without potential (i.e. $\mathbf{P} = 0$). Then the multiplicity of the smallest eigenvalue $\lambda_1$ of $\mathbf{W}(G)$ is equal to the number of components of $G$. In particular, $\lambda_1$ is simple if and only if $G$ is connected.*

*Proof.* Assume $G$ is the disjoint sum of connected components $H_1, \ldots, H_k$. Denote by $f_i$ the characteristic function of $V(H_i)$, i.e. $f(v) = 1$ if $v \in V(H_i)$ and 0 otherwise. Obviously, $\mathbf{M}(G)f_i = 0$. Since $f_1, \ldots, f_k$ are linearly independent, the multiplicity of eigenvalue 0 is at least $k$.

Conversely, if $f$ is an eigenfunction of eigenvalue 0, then by (2.6) $f$ must be constant on each edge of $G$ and hence on each component $H_i$. Therefore $f$ is a linear combination of the characteristic functions $f_i$.    □

We assume throughout this book that all graphs are connected unless stated otherwise explicitly.

## 2.2 Weighted Graphs

We have introduced Laplacian and generalized Laplacian matrices on simple unweighted graphs. However, it is straightforward to generalize these concepts to *weighted graphs*. Let $w_{xy} > 0$ denote the weight for edge $xy$; we set $w_{xy} = 0$ if $x$ and $y$ are not adjacent. Then we can define the Laplacian $\mathbf{L}_w$ as

$$(\mathbf{L}_w f)(x) = \sum_{y \sim x} w_{xy}(f(x) - f(y)) . \qquad (2.8)$$

Obviously this is a special case of (2.5) with $-M_{xy} = w_{xy}$ and $p(x) = 0$. Thus $\mathbf{L}_w$ can be seen as a generalized Laplacian on the corresponding unweighted graph (where two vertices $x$ and $y$ are adjacent if and only if $w_{xy} > 0$). Thus without loss of generality we will restrict our interest to generalized Laplacian on unweighted graphs.

## 2.3 The Rayleigh Quotient

The *Rayleigh quotient* $\mathcal{R}_{\mathbf{M}}(f)$ of a function $f\colon V \to \mathbb{R}$ with respect to a generalized Laplacian $\mathbf{M}$ is defined as the fraction

$$\mathcal{R}_{\mathbf{M}}(f) = \frac{\langle f, \mathbf{M}f \rangle}{\langle f, f \rangle} \, . \tag{2.9}$$

For the graph Laplacian $\mathbf{L}$ this can equivalently be written as

$$\mathcal{R}_{\mathbf{L}}(f) = \frac{\sum_{xy \in E}(f(x) - f(y))^2}{\sum_{x \in V} f(x)^2} \, .$$

The Rayleigh quotient plays a crucial rôle in our investigations. Its importance is based on the following fundamental theorem from spectral theory for symmetric matrices (which we restate here for graph Laplacians), see e.g. [100].

**Proposition 2.4 (Spectral Decomposition).** *For a generalized Laplacian* $\mathbf{M}$ *for a graph* $G$ *there exists an orthonormal basis of the* $\mathbb{R}^n$ *that consists of eigenfunctions* $f_1, \ldots, f_n$ *corresponding to the eigenvalues* $\lambda_1, \ldots, \lambda_n$. *Moreover, for every function* $g\colon V \to \mathbb{R}$ *we find*

$$\mathbf{M}g = \sum_{i=1}^{n} \lambda_i \langle g, f_i \rangle f_i$$

*and for the quadratic form,*

$$\langle g, \mathbf{M}g \rangle = \sum_{i=1}^{n} \lambda_i \langle g, f_i \rangle^2 \, .$$

As an immediate consequence we have the following corollary.

**Corollary 2.5.** *Let* $f_1, \ldots, f_n$ *denote orthogonal eigenfunctions corresponding to the eigenvalues* $\lambda_1 \le \lambda_2 \le \cdots \le \lambda_n$ *of a generalized Laplacian* $\mathbf{M}$. *Let* $F_i = \{f_1, \ldots, f_i\}$ *be the set of the first* $i$ *eigenfunctions and* $F_i^{\perp}$ *its orthogonal complement. Then*

$$\lambda_k = \min_{g \in F_{k-1}^{\perp}} \mathcal{R}_{\mathbf{M}}(g) = \min_{g \in F_{k-1}^{\perp}} \frac{\langle g, \mathbf{M}g \rangle}{\langle g, g \rangle} \, .$$

*Moreover,* $\mathcal{R}_{\mathbf{M}}(g) = \lambda_k$ *for some* $g \in F_{k-1}^{\perp}$ *if and only if* $g$ *is an eigenfunction corresponding to* $\lambda_k$.

*Proof.* Every function $g \in F_{k-1}^{\perp}$ can be written as $g = \sum_{i=k}^{n} a_i f_i$. Hence $\mathcal{R}_{\mathbf{M}}(g) = \sum_{i=k}^{n} \lambda_i a_i^2 / \sum_{i=k}^{n} a_i^2 \ge \sum_{i=k}^{n} \lambda_k a_i^2 / \sum_{i=k}^{n} a_i^2 = \lambda_k$ and equality holds if and only if all terms with eigenvalues $\lambda_i > \lambda_k$ vanish. Thus the result follows. $\qquad\square$

**Corollary 2.6 (Minimax-Theorem).** *Let $\mathcal{W}_k$ and $\mathcal{W}_k^{\perp}$ denote the sets of subspaces of $\mathbb{R}^n$ of dimension at least $k$ and of codimension at most $k$, respectively. Then*

$$\lambda_k = \min_{W \in \mathcal{W}_k} \max_{0 \neq g \in W} \frac{\langle g, \mathbf{M}g \rangle}{\langle g, g \rangle} = \max_{W \in \mathcal{W}_{k-1}^{\perp}} \min_{0 \neq g \in W} \frac{\langle g, \mathbf{M}g \rangle}{\langle g, g \rangle}$$

*Proof.* Every function $g$ can be written as $g = \sum_{i=1}^n a_i f_i$ for some $a_i$ where $\{f_1, \ldots, f_n\}$ is the orthonormal basis of eigenfunctions from Prop. 2.4. Hence $\mathcal{R}_{\mathbf{M}}(g) = \frac{\langle g, \mathbf{M}g \rangle}{\langle g, g \rangle} = \sum_{i=1}^n a_i^2 \lambda_i / \sum_{i=1}^n a_i^2$. Then for every $W \in \mathcal{W}_k$ we can find some $g \in W$ where $a_1 = \ldots a_{k-1} = 0$ and thus $\sup_{g \in W} \mathcal{R}_{\mathbf{M}}(g) \geq \sup_{g \in W, a_1 = \ldots = a_{k-1} = 0} \sum_{i=k}^n a_i^2 \lambda_i / \sum_{i=k}^n a_i^2 \geq \lambda_k$. Consequently,

$$\inf_{W \in \mathcal{W}_k} \sup_{g \in W} \mathcal{R}_{\mathbf{M}}(g) \geq \lambda_k \ .$$

Equality holds if $W$ is the subspace that is spanned by the first $k$ eigenfunctions. Thus the first equality follows. The second equality is shown analogously. $\square$

## 2.4 Calculus on Graphs

Friedman and Tillich [76, 77] developed a *Calculus on Graphs* where ideas for motivating the discrete Dirichlet matrix [75] are extended to a more general setting; see Sect. 1.5 for a more detailed description.

The *geometric realization* of a graph $G(V, E)$ is the metric space $\mathcal{G}$ consisting of $V$ and arcs of length 1 glued between $u$ and $v$ for every edge $e = uv \in E$. For weighted graphs these arcs have length $1/w_{uv}$. This definition of the arc lengths needs some explanation. Setting the length of such arcs to the reciprocal of weights of the corresponding edge is motivated by the application of graphs in physical models (see e.g. Hückel theory in Sect. 1.6) or in numerical approximations of the continuous operators (see e.g. Sect. 1.2). Shorter distances between the nodes (i.e., smaller arc lengths) result in stronger coupling in these systems and hence are modeled by higher weights for these connections.

We define two measures on $\mathcal{G}$ (and $G$). A *vertex measure*, $\mu_V$, is supported on the vertex set $V$ with $\mu_V(v) > 0$ for all $v \in V$; and an *edge measure* $\mu_E$, supported on the union of arcs of $\mathcal{G}$, with $\mu_E(v) = 0$ for all $v \in V$ and whose restriction to any open subinterval of an edge (arc) $e \in E$ is its Lebesgue measure times a constant $a_e > 0$. In our setup we have the measures $\mu_V(v) = 1$ and $a_e = 1$ (which are called *traditional* in [76]). Hence for any graph $G$, $\mu_V(G) = |V|$ and $\mu_E(G) = |E|$ (or $\sum_{e \in E} 1/w_e$ in case of a weighted graph).

Let $\mathcal{S}$ denote the set of all continuous functions on $\mathcal{G}$ which are differentiable on $\mathcal{G} \setminus V$. Then we introduce a Laplacian operator $\mathcal{L}(\mathcal{G})$ by the Rayleigh quotient for functions $f \in \mathcal{S}$ given as

$$\mathcal{R}_{\mathcal{L}}(f) = \frac{\int_{\mathcal{G}} |\nabla f|^2 d\mu_E}{\int_{\mathcal{G}} |f|^2 d\mu_V}.$$

The operator $\mathcal{L}(\mathcal{G})$ can be seen as the continuous version of the corresponding graph Laplacian $\mathbf{L}(G)$. On $\mathcal{G}$ we can avoid the problems that arise from the discreteness of our situation. Many concepts in analysis translate almost immediately to this setting. For example, nodal domains (Sect. 3.1) of an (eigen-) function $f$ are separated by points in $\mathcal{G}$ where $f$ vanishes; in opposition to the traditional setting where such points need not exist as $f$ is supported on $V$ only (see Sect. 3.1).

These two concepts, $\mathbf{L}(G)$ and $\mathcal{L}(\mathcal{G})$, coincide [75, 76]. The Rayleigh quotient $\mathcal{R}_{\mathcal{L}}(f)$ is minimized if and only if $f \in \mathcal{S}$ is an edgewise linear function, i.e. a function whose restriction to an edge is linear. The eigenvalues and eigenfunctions of $\mathcal{L}(\mathcal{G})$ exist and are those of $\mathbf{L}(G)$, i.e. the restrictions of the $\mathcal{L}(\mathcal{G})$-eigenfunctions to $V$ are the graph Laplacian eigenfunctions.

In this setting the motivation for the Dirichlet operator, introduced in Sect. 1.5, is obvious: Restrict $\mathcal{S}$ to $\{f \in \mathcal{S}: f(v) = 0 \text{ for all } v \in \partial V\}$. We then have the following analog to eigenfunctions of the classical Laplace-Beltrami operator. If $G_1$ and $G_2$ are graphs with boundary, then we say that $G_2$ is an extension of $G_1$, written $G_1 \subseteq G_2$, if there exists an isometric embedding of the realization of $G_1$ into $G_2$ which preserves the degree of each interior vertex. If $G_1$ and $G_2$ are connected graphs and the above embedding is not onto, we say that $G_2$ is a strict extension, $G_1 \subset G_2$.

**Proposition 2.7 ([75]).** *Let $\lambda^\circ(G)$ denote the first Dirichlet eigenvalue. Then the following holds:*

*(1) $\lambda^\circ(G)$ is continuous as a function of $G$ in the metric*
   $\rho(G, G') = \mu_E(G - G') + \mu_E(G' - G)$.
*(2) $\lambda^\circ(G)$ is monotone in $G$, i.e., if $G \subset G'$ then $\lambda^\circ(G) > \lambda^\circ(G')$.*

## 2.5 Basic Properties of Eigenfunctions

As we have already seen the graph Laplacian is a nonnegative operator. If $G$ is a connected graph with $n$ vertices then the constant function $\mathbf{1}: x \mapsto 1$ is the unique eigenfunction with eigenvalue 0, $\mathbf{L1} = 0$ (for a proof see Cor. 2.23). Each eigenfunction of an eigenvalue greater than 0 is orthogonal to $\mathbf{1}$ by Prop. 2.4. Thus there are at least two vertices with values of opposite sign, and of course $\sum_{x \in V} f(x) = 0$. For vertices where an eigenfunction vanishes we have the following important property which holds for every generalized Laplacian.

**Lemma 2.8.** *Let $f$ be an eigenfunction of $\mathbf{M}(G)$ with a zero vertex $z$, i.e., a vertex where $f$ vanishes, $f(z) = 0$. Then $\sum_{y \sim z} M_{yz} f(y) = 0$. Moreover, either all neighbors of the zero vertex $z$ are zero vertices themselves, or $z$ is adjacent to vertices of both strict signs.*

*Proof.* $0 = f(z) = \sum_{y \in V} M_{yz} f(y) = \sum_{y \sim z} M_{yz} f(y) + M_{zz} f(z) = \sum_{y \sim z} M_{yz} f(y).$

□

The next property can be interpreted as a discrete analog of the *maximum principle* for the Laplace operator. We say that $x$ is a *local maximum* of a function $f$ if $f(x) \geq f(y)$ for all $y \sim x$ and $f(x) > f(z)$ for at least one $z \sim x$. A *local minimum* is defined analogously.

**Theorem 2.9 ([81, 90]).** *An eigenfunction $f$ of a graph Laplacian $\mathbf{L}(G)$ cannot have a nonnegative local minimum or a nonpositive local maximum.*

*Proof.* Suppose $x$ is a local minimum of $f$ with $f(x) \geq 0$. Then $\sum_{y \sim x}[f(x) - f(y)] < 0$ and thus by (2.2), $0 \leq \lambda f(x) = (\mathbf{L}f)(x) = \sum_{y \sim x}[f(x) - f(y)] < 0$, a contradiction.                                                                        □

*Remark 2.10.* This theorem analogously holds for generalized Laplacians without a potential $p(x)$ in (2.5). However, if $p(x) \neq 0$ for some vertices then it might fail. For example, consider a simple path $P_3$, with generalized Laplacian

$$\mathbf{M} = \begin{pmatrix} 3 & -1 & -1 \\ -1 & 1 & 0 \\ -1 & 0 & 1 \end{pmatrix}.$$

Then $\lambda_1 = 2 - \sqrt{3}$ has an eigenfunction with a positive minimum on the second vertex.

Merris [134] considers several "eigenfunction principles" for the graph Laplacian. In the following we review some of them.

**Theorem 2.11 ([134]).** *Let $G$ be a graph with $n$ vertices. If $0 \neq \lambda < n$ is an eigenvalue of $\mathbf{L}(G)$, then any eigenfunction affording $\lambda$ takes the value 0 on every vertex of degree $n - 1$.*

*Proof.* Let $v$ be a vertex of degree $n-1$. $(\mathbf{L}f)(v) = (n-1) f(v) - \sum_{x \neq v} f(x) = \lambda f(v)$, hence $n f(v) = \lambda f(v)$ and $f(v) = 0$.                                              □

**Theorem 2.12 ([134]).** *Let $\lambda$ be an eigenvalue of $\mathbf{L}(G)$ afforded by eigenfunction $f$. If $f(u) = f(v)$, then $\lambda$ is an eigenvalue of $\mathbf{L}(G')$ afforded by $f$, where $G'$ is the graph obtained from $G$ by deleting or adding the edge $e = uv$ depending on whether or not $e = uv$ is an edge of $G$.*

The *reduced graph* $G\{W\}$ is obtained from $G$ by deleting all vertices in $V \setminus W$ that are not adjacent to a vertex of $W$ and subsequent deletion of any remaining edges that are not incident with a vertex of $W$.

**Theorem 2.13 ([134]).** *For a graph $G(V, E)$ fix a nonempty subset $W$ of $V$. Suppose $f$ is an eigenfunction of the reduced graph $G\{W\}$ that affords $\lambda$ and is supported by $W$ in the sense that if $f(u) \neq 0$, then $u \in W$. Then the extension $f'$ with $f'(v) = f(v)$ for $v \in W$ and $f'(v) = 0$ otherwise is an eigenfunction of $G$ affording $\lambda$.*

**Theorem 2.14 ([134]).** *Let f be an eigenfunction affording λ of a graph G with n vertices. Let $N_v$ be the set of neighbors of v. Suppose $f(u) = f(v) = 0$, where $N_u \cap N_v = \emptyset$. Let G′ be the graph on $n - 1$ vertices obtained by coalescing u and v into a single vertex, which is adjacent in G′ precisely to those vertices that are adjacent in G to u or to v. The function f′ obtained by restricting f to $V(G) \setminus \{u\}$ is an eigenfunction of G′ affording λ.*

If G is a regular graph, then the eigenvalues of the Laplacian are determined by the eigenvalues of the adjacency matrix.

**Proposition 2.15.** *Let G be a k-regular graph. If the adjacency matrix $\mathbf{A}(G)$ has eigenvalues $\lambda_1, \ldots, \lambda_n$, then the Laplacian $\mathbf{L}(G)$ has eigenvalues $k - \lambda_1, \ldots, k - \lambda_n$.*

*Proof.* If G is k-regular, then $\mathbf{L}(G) = \mathbf{D}(G) - \mathbf{A}(G) = k\mathbf{I} - \mathbf{A}$. Thus every eigenfunction of $\mathbf{A}$ with eigenvalue λ is an eigenfunction of $\mathbf{L}(G)$ with eigenvalue $k - \lambda$.    □

The next well-known result describes the relation between the Laplacian spectrum of G and the Laplacian spectrum of its complement $G^c$. The matrix $\mathbf{J}$ is the $n \times n$ matrix each of whose entries are 1.

**Theorem 2.16.** *If G is a graph with n vertices and f is an eigenfunction of $\mathbf{L}(G)$ with eigenvalue $\lambda \neq 0$, then f is an eigenfunction of $\mathbf{L}(G^c)$ with eigenvalue $n - \lambda$.*

*Proof.* We start observing that $\mathbf{L}(G) + \mathbf{L}(G^c) = n\mathbf{I} - \mathbf{J}$ and $\mathbf{J}f = 0$ as f is orthogonal to the constant function $\mathbf{1}$. Then,

$$nf = (n\mathbf{I} - \mathbf{J})f = \mathbf{L}(G)f + \mathbf{L}(G^c)f = \lambda f + \mathbf{L}(G^c)f .$$

Therefore, $\mathbf{L}(G^c)f = (n - \lambda)f$.    □

## 2.6 Graph Automorphisms and Eigenfunctions

It is sometimes possible to infer directly from the graph structure at which vertices some or all eigenfunctions of $\mathbf{L}(G)$ vanish. Theorem 2.11 is an example. Symmetry properties of G are particularly useful for this purpose.

An automorphism of a graph G is a permutation of its vertex set $V(G)$ that maps edges onto edges and nonedges onto nonedges. The set of all automorphisms of G forms a group. We denote this *automorphism group* of G by Aut(G). For an $X \in \text{Aut}(G)$ and a given eigenfunction f we define the function $Xf$ by

$$Xf(v) = f(X(v)) .$$

Moreover

$$V_X = \{v \in V \colon X(v) = v\} \qquad \text{and} \qquad O_X(v) = \{X^k(v) \colon k \in \mathbb{Z}\}$$

denote the set of vertices that are fixed under the action of $X$ and the orbit of the vertex $v$ under the action of $X$, respectively.

**Lemma 2.17.** *Let $X \in \mathrm{Aut}(G)$ for some graph $G$. If $f$ is an eigenfunction of $\mathbf{L}(G)$ corresponding to eigenvalue $\lambda$, then $Xf$ is also an eigenfunction of $\lambda$.*

*Proof.* $\mathbf{L}(Xf)(v) = \sum_{w \sim v}(Xf(v) - Xf(w)) = \sum_{w \sim v}(f(X(v)) - f(X(w))) = \sum_{w \sim X(v)}(f(X(v)) - f(w)) = \mathbf{L}f(X(v)) = \lambda f(X(v)) = \lambda X f(v).$ $\qquad\square$

**Theorem 2.18.** *For an eigenfunction $f$ and an automorphism $X \in \mathrm{Aut}(G)$ one of the following three cases holds:*

*(1) $Xf = f$. In particular, $f$ is constant on every orbit $O_X(v)$.*
*(2) $Xf = -f$, and $f$ vanishes on all orbits of odd size. In particular, $f$ vanishes on the fixed points $V_X$. Moreover, there must be an orbit of even size.*
*(3) $Xf$ and $f$ are linearly independent, and consequently $\lambda$ is an eigenvalue of multiplicity greater than one.*

*Proof.* Let $s$ denote the size of the orbit $O_X(v)$ of vertex $v$ ($s = 1$ if $v \in V_X$), i.e., $X^s v = v$. Assume $Xf = \alpha f$ for some $\alpha \in \mathbb{R}$. Then we find $f(v) = f(X^s v) = Xf(X^{s-1}v) = \alpha f(X^{s-1}v) = \cdots = \alpha^s f(v)$. Thus $f(v) = 0$ and $f$ vanishes on the orbit of $v$, or $\alpha^s = 1$ and hence $\alpha = 1$ (case (1)), or $\alpha = -1$ (case (2)). Obviously if $f(v) = X^s f(v) = (-1)^s f(v)$ then $f$ vanishes on all orbits of odd size $s$ and there must be an orbit of even size since otherwise $f$ would be identical to zero. Another immediate consequence of these considerations is that when neither (1) nor (2) holds, then $Xf$ and $f$ are linearly independent (case (3)). $\qquad\square$

**Theorem 2.19.** *Let $X \in \mathrm{Aut}(G)$ and let $f_1$ and $f_2$ be Laplacian eigenfunctions of $G$ with properties (1) and (2) of Thm. 2.18, respectively. Then $f_1$ and $f_2$ are orthogonal, i.e., $\langle f_1, f_2 \rangle = 0$.*

*Proof.* Since $X$ is a permutation operator on $V$, we have $X^t X = I$. Thus we find $\langle f_1, f_2 \rangle = \langle X^t X f_1, f_2 \rangle = \langle X f_1, X f_2 \rangle = \langle f_1, -f_2 \rangle = -\langle f_1, f_2 \rangle$ and the proposition follows. $\qquad\square$

## 2.7 Quasi-Abelian Cayley Graphs

In highly symmetric graphs one can expect a close connection between eigenfunctions of the graph Laplacian and group-theoretic properties. We exploit this connection here to derive explicit expressions for the eigenfunctions of the graph Laplacian of a class of highly symmetric graphs.

Let $\mathsf{G}$ be a finite group and let $S$ be a symmetric set of generators of $\mathsf{G}$, i.e., $\langle S \rangle = \mathsf{G}$, $S = S^{-1}$, and $\imath \notin S$, where $\imath$ is the identity of $\mathsf{G}$. A graph

$\Gamma(\mathsf{G},S)$ with vertex set $\mathsf{G}$ and edges $\{s,t\}$ if and only if $t^{-1}s \in S$ is called a *Cayley graph*. A Cayley graph $\Gamma(\mathsf{G},S)$ is called *quasi-Abelian* if $S$ is the union of some conjugacy classes of $\mathsf{G}$.

Cayley graphs are vertex transitive and hence regular. The characteristic function of $S$ will be denoted by $\Theta : \mathsf{G} \to \{0,1\}$. Clearly, a Cayley graph on a commutative group is quasi-Abelian, because in this case each group element forms its own conjugacy class. Some interesting properties of quasi-Abelian Cayley graphs are discussed in [172, 179].

In the case of Cayley graphs we have to distinguish between the "Fourier series expansion" (1.5) with respect to the Laplacian matrix of the graph $\Gamma(\mathsf{G},S)$, and the representation theoretical Fourier transformation on the group $\mathsf{G}$ itself. It should not come as a surprise that there is an intimate connection between these two. In fact, the connection between the algebraic properties of $\Gamma(\mathsf{G},S)$ and the representation theory of the underlying group $\mathsf{G}$ derives from the following simple facts: The *regular representation* $\boldsymbol{\rho}_{\mathrm{reg}}$ of $\mathsf{G}$ is defined by

$$\boldsymbol{\rho}_{\mathrm{reg}}(s)f(t) = f(s^{-1}t)$$

for any $f : \mathsf{G} \longrightarrow \mathbb{C}$. Substituting $\Theta$ for $f$ we find $\boldsymbol{\rho}_{\mathrm{reg}}(s)\Theta(t) = \Theta(s^{-1}t) = 1$ if $\{t,s\}$ is an edge of $\Gamma(\mathsf{G},S)$ and 0 otherwise. Thus we may write the adjacency matrix $\mathbf{A}(\mathsf{G},S)$ of $\Gamma(\mathsf{G},S)$ in the form

$$\mathbf{A}(\mathsf{G},S) = \sum_{s \in S} \boldsymbol{\rho}_{\mathrm{reg}}(s) .$$

For any function $f : \mathsf{G} \to \mathbb{C}$ and any matrix representation $\varrho = \{\boldsymbol{\rho}(s)\}_{s \in \mathsf{G}}$ of $\mathsf{G}$ we call the matrix sum

$$\widehat{f}(\varrho) = \sum_{x \in \mathsf{G}} f(x)\boldsymbol{\rho}(x)$$

the (group theoretic) *Fourier Transform* of $f$ at $\varrho$. Consider a complete set $\{\varrho^1, \ldots, \varrho^h\}$ of inequivalent irreducible matrix representations of $\mathsf{G}$. Let $d_k$ denote the dimension of $\varrho^k$. Then

$$f(s) = \frac{1}{|\mathsf{G}|} \sum_{k=1}^{h} d_k \operatorname{tr}(\boldsymbol{\rho}^k(s^{-1})) \, \widehat{f}(\varrho^k)$$

inverts the Fourier transform.

Following e.g. [55, Sect. 8A] we assume that the irreducible representations $\varrho^k$ are unitary, i.e., that $\boldsymbol{\rho}^k(t)^* = \boldsymbol{\rho}^k(t^{-1})$ and introduce

$$\tilde{\rho}_{ij}^k(s) := \sqrt{d_k}\rho_{ji}^k(s^{-1}) .$$

These functions are orthonormal w.r.t. the scalar product

$$\langle \varphi, \psi \rangle = \frac{1}{|\mathsf{G}|} \sum_{s \in |\mathsf{G}|} \varphi(s)\psi^*(s)$$

and form a new basis for the vector space of functions of $G$. Now we are in the position to state the main result of this section.

**Theorem 2.20 ([151]).** *Let $\Gamma(G, S)$ be a quasi-Abelian Cayley graph with a finite group $G$.*

*(i) The function $\varepsilon_{ij}^k : G \to \mathbb{C}$ defined as*

$$\varepsilon_{ij}^k(u) = \frac{1}{\sqrt{|G|}} \tilde{\rho}_{ij}^k(u) = \sqrt{\frac{d_k}{|G|}} \rho_{ij}^k(u^{-1})$$

*is a normalized eigenfunction of $\mathbf{L}(\Gamma)$ with eigenvalue*

$$\lambda_k = |S| - \frac{1}{d_k} \sum_{s \in S} \chi_k(s)$$

*where $\chi_k(s) = \mathrm{tr}(\boldsymbol{\rho}^k(s))$ is the character of $\varrho^k$ at $s$.*

*(ii) All quasi-Abelian Cayley graphs on $G$ have a common basis of eigenfunctions and hence their Laplacian matrices commute.*

*Proof.* (i) We verify by explicit computation that $\tilde{\rho}_{ij}^k$ is an eigenfunction of the adjacency matrix:

$$\sum_{u \in G} \mathbf{A}_{vu} \tilde{\rho}_{ij}^k(u) = \sum_{u \in G} \Theta(vu^{-1}) \tilde{\rho}_{ij}^k(u)$$

$$= \sum_{u \in G} \left\{ \frac{1}{|G|} \sum_{r,s,t} \sqrt{d_r} \widehat{\Theta}_{ts}(\rho^r) \tilde{\rho}_{st}^r(uv^{-1}) \right\} \tilde{\rho}_{ij}^k(u)$$

$$= \sum_{u \in G} \frac{1}{|G|} \sum_{r,s,t} \widehat{\Theta}_{ts}(\varrho^r) \sum_{y} \tilde{\rho}_{ys}^{r*}(u) \tilde{\rho}_{yt}^r(v) \tilde{\rho}_{ij}^k(u)$$

$$= \sum_{r,s,t} \widehat{\Theta}_{ts}(\varrho^r) \sum_{y} \tilde{\rho}_{yt}^r(v) \frac{1}{|G|} \sum_{u \in G} \tilde{\rho}_{ij}^k(u) \tilde{\rho}_{ys}^{r*}(u)$$

$$= \sum_{r,s,t} \widehat{\Theta}_{ts}(\varrho^r) \sum_{y} \tilde{\rho}_{yt}^r(v) \delta_{kr} \delta_{iy} \delta_{js} = \sum_{t} \widehat{\Theta}_{tj}(\varrho^k) \tilde{\rho}_{it}^k(v) \ .$$

Here we have used that $\boldsymbol{\rho}^k(st^{-1}) = \boldsymbol{\rho}^k(s)\boldsymbol{\rho}^k(t^{-1}) = \boldsymbol{\rho}^k(s)\boldsymbol{\rho}^{k*}(t)$ translates to

$$\sqrt{d_r} \tilde{\rho}_{st}^r(vu^{-1}) = \sum_{y=1}^{h} \tilde{\rho}_{ys}^{r*}(u) \tilde{\rho}_{yt}^r(v) \ .$$

Next we use the fact that $\Theta$ is a class function. Hence its Fourier transform is diagonal

$$\widehat{\Theta}(\rho^k) = \frac{1}{d_k} \sum_{s \in S} \chi_k(s) \mathbf{I}_{d_k}$$

where $\chi_k(s) = \mathrm{tr}(\boldsymbol{\rho}^k(s))$ is the character of the representation $\varrho^k$ at $s$. We have therefore

$$\sum_{u \in \mathsf{G}} \mathbf{A}_{vu} \tilde{\rho}_{ij}^k(u) = \sum_t \frac{1}{d_k} \sum_{s \in S} \chi_k(s) \delta_{tj} \tilde{\rho}_{it}^k(v) = \frac{1}{d_k} \sum_{s \in S} \chi_k(s) \times \tilde{\rho}_{ij}^k(v) \ .$$

Changing the normalizations back to the standard scalar product of $\mathbb{C}$ and using $\mathbf{L} = |S|\mathbf{I} - \mathbf{A}$ leads to claim (i) of the theorem.

(ii) We have just shown that $\{\tilde{\rho}^{ij}\}$ is an orthonormal basis of eigenfunctions of $\mathbf{L}$ whenever $S$ is the union of conjugacy classes of $\mathsf{G}$. Thus the Laplacian matrices of all quasi-Abelian Cayley graphs of the group $\mathsf{G}$ share a common orthonormal basis of eigenfunctions. Since the graph Laplacians are symmetric matrices, they commute under these circumstances.   □

Theorem 2.20 generalizes the following well known result for Abelian Cayley graphs which is discussed e.g. by Lovász [128]:

**Corollary 2.21.** *Let $\mathsf{G}$ be a commutative group, and let $S$ be a symmetric set of generators of $\mathsf{G}$. Then the irreducible characters $\chi_k$ of $\mathsf{G}$ are eigenfunctions of $\mathbf{A}(\mathsf{G}, S)$ with corresponding eigenvalue $\Lambda_k = \sum_{s \in S} \chi_k(s)$.*

## 2.8 The Perron-Frobenius Theorem

Let $\mathbf{A}$ be an $n \times n$ real symmetric matrix. Analogously to the generalized Laplacians we can associate a graph $G$ such that two vertices $u$ and $v$ are connected by an edge if and only if $A_{uv} \neq 0$. Then $\mathbf{A}$ is called *irreducible* if its underlying graph is connected[1].

**Theorem 2.22 (Perron-Frobenius Theorem).** *Let $\mathbf{A}$ and $\mathbf{B}$ be real symmetric irreducible nonnegative $n \times n$ matrices. Then*

*(i) the spectral radius $\rho(\mathbf{A})$ is a simple eigenvalue of $\mathbf{A}$. If $\mathbf{x}$ is an eigenfunction for $\rho(\mathbf{A})$, then no entries of $x$ are zero, and all have the same sign.*

*(ii) If moreover $\mathbf{A} - \mathbf{B}$ is nonnegative, then $\rho(\mathbf{B}) \leq \rho(\mathbf{A})$, with equality if and only if $\mathbf{B} = \mathbf{A}$.*

For a proof see, e.g., [100].

We can apply this theorem to get a statement about the smallest eigenvalue $\lambda_1$ and its eigenfunctions of a generalized Laplacian of $G$.

**Corollary 2.23.** *Let $G$ be a connected graph with a generalized Laplacian $\mathbf{M}$. Then the smallest eigenvalue $\lambda_1$ of $\mathbf{M}$ is simple and the corresponding eigenfunction can be taken to have all entries positive.*

---

[1] A nonsymmetric matrix is called irreducible if the corresponding graph is strongly connected, i.e., if, for all $u, v \in V$, there is a *directed* path from $u$ to $v$. The Perron-Frobenius Theorem then holds as well.

*Proof.* We use an argument of Godsil and Royle [85]. If $\mathbf{M}$ is a generalized Laplacian of $G$, then for any $c$, the matrix $\mathbf{M} - c\mathbf{I}$ is a generalized Laplacian of $G$ with the same eigenfunctions as $\mathbf{M}$. We choose a constant $c$ such that all diagonal entries of $\mathbf{M} - c\mathbf{I}$ are nonpositive. As a consequence of the Perron-Frobenius Theorem, the largest eigenvalue of $-\mathbf{M} + c\mathbf{I}$ is simple and the associated eigenfunction may be taken to have only positive entries. $\quad\Box$

A positive eigenfunction to the smallest eigenvalue $\lambda_1$ of $\mathbf{M}$ of a connected graph is called a *Perron vector* of $G$.

# 3

# Eigenfunctions and Nodal Domains (From Perron Frobenius to Courant's Nodal Domain Theorem)

In the previous chapter we have seen that (due to the Perron-Frobenius Theorem) the eigenfunctions of the first eigenvalue $\lambda_1$ have all entries positive (or negative) for a generalized Laplacian matrix $\mathbf{M}$ of a connected graph $G$. Fiedler [67] has shown that for eigenfunctions of the smallest nonzero eigenvalue of a graph the subgraph induced by nonpositive vertices (i.e., vertices with nonpositive function values) and the subgraph induced by nonnegative vertices are both connected. In other words, an eigenfunction of the second eigenvalue has exactly two *weak nodal domains* (also called *weak sign graphs*).

These two observations remind us on Courant's celebrated Nodal Domain Theorem for elliptic operators on manifolds. Courant [44, Chap. 6, §6] stated a general theorem about the "nodes" of an eigenfunction: *Given the self-adjoint second order differential equation $L[u] + \lambda \rho u = 0$ ($\rho > 0$) for a domain $G$ with arbitrary homogeneous boundary conditions; if its eigenfunctions are ordered according to increasing eigenvalues, then the nodes of the $n$-th eigenfunction $u_n$ divide the domain into no more than $n$ subdomains. No assumptions are made about the number of independent variables.* Courant's "subdomains" have since then become known as *nodal domains*, see e.g. [29, 31]. In this chapter we see that the eigenfunctions of discrete Laplace operators have similar properties.

## 3.1 Courant's Nodal Domain Theorem

In the context of manifolds, "nodes" are points where the eigenfunction $u$ vanishes, i.e., the nodal set of $u$ is $\{x|u(x) = 0\}$. The nodal sets themselves are known to be of zero Lebesgue measure and of codimension 1, [31, 104]. The term "nodal domain" refers to the connected components of the complement of the nodal set, i.e., to the components of $\{x: u(x) \neq 0\}$, which are bounded by nodal sets. This terminology is now well-established in the PDE literature. Of course, it is not well suited for graphs: A discrete eigenfunction of a graph Laplacian is defined only on the vertex set $V$ of a graph $G$ and thus, contrary to the situation on a manifold, it may change from positive to negative sign without passing through a zero.

The discrete analog of a "nodal domain" is thus a maximal connected induced subgraph consisting entirely of positive and negative vertices, which would more appropriately be called a *sign graph*. Nevertheless, we will use the traditional notion of *nodal domains* in this book to emphasize the analogy of many of the results with the case of PDEs on manifolds.

The examples in Figs. 3.1 and 3.2 show, however, that we cannot adopt the original continuous version without modification. In the star graph in Fig. 3.1 we count 4 connected components in which the eigenfunction of the second eigenvalue is strictly positive or strictly negative, respectively. The eigenfunctions of the Petersen graph serve as a second counterexample (Fig. 3.2). This important rôle of zero vertices was already observed by Powers [147] who extended some of Fiedler's results (see Cor. 3.3 below) for the adjacency matrix. We are forced to distinguish between two versions of nodal domains.

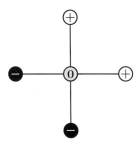

**Fig. 3.1.** Sign pattern of a particular eigenfunction $f$ of the second eigenvalue of the Laplacian of a star $G$: $\mathfrak{S}(f) = 4 > 2$ and $\mathfrak{W}(f) = 2$. The numerical values of the eigenfunctions can be found in Appendix B.

A *positive (negative) strong nodal domain* of a function $f$ on $V(G)$ is a maximal connected induced subgraph of $G$ on vertices $v \in V$ with $f(v) > 0$ ($f(v) < 0$). In contrast, a *positive (negative) weak nodal domain* of a function $f$ on $V(G)$ is a maximal connected induced subgraph of $G$ on vertices $v \in V$ with $f(v) \geq 0$ ($f(v) \leq 0$) that contains at least one nonzero vertex. In the following we will be interested in the number of strong and weak nodal domains of a function $f$ which we denote by $\mathfrak{S}(f)$ and $\mathfrak{W}(f)$, respectively. Obviously, $\mathfrak{W}(f) \leq \mathfrak{S}(f)$. Figures 3.1 and 3.2 show two examples.

The obvious difference between the definitions of strong and weak nodal domains is the rôle of *zero vertices*, i.e. vertices where the function $f$ vanishes. While such vertices separate positive (or negative) strong nodal domains, they join weak nodal domains. In fact, each zero vertex simultaneously belongs to exactly one weak positive nodal domain and exactly one weak negative nodal domain. If two different weak nodal domains $D_1$ and $D_2$ overlap, then they must have opposite signs except on zero vertices. In the following we will only consider nodal domains of an eigenfunction of a generalized Laplacian.

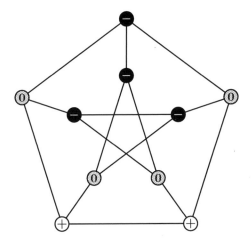

**Fig. 3.2.** Sign pattern of a particular eigenfunction $f$ of the second eigenvalue of the Laplacian of the Petersen graph: $\mathfrak{S}(f) = 3 > 2$ and $\mathfrak{W}(f) = 2$.

We focus our attention on the $k$-th eigenvalue $\lambda_k$ with multiplicity $r$ of a generalized Laplacian $\mathbf{M}$. We assume throughout this and the following chapters that the eigenvalues are labeled in ascending order starting with 1, so that

$$\lambda_1 \leq \cdots \leq \lambda_{k-1} < \lambda_k = \lambda_{k+1} = \cdots = \lambda_{k+r-1} < \lambda_{k+r} \leq \cdots \leq \lambda_n.$$

We are now in the position to formulate discrete versions of Courant's Nodal Domain Theorem.

**Theorem 3.1 (Discrete Nodal Domain Theorem, [49]).** *Let $\mathbf{M}$ be a generalized Laplacian of a connected graph with $n$ vertices. Then any eigenfunction $f_k$ corresponding to the $k$-th eigenvalue $\lambda_k$ with multiplicity $r$ has at most $k$ weak nodal domains and $k + r - 1$ strong nodal domains:*

$$\mathfrak{W}(f_k) \leq k \qquad and \qquad \mathfrak{S}(f_k) \leq k + r - 1 \,. \tag{3.1}$$

Various versions of the nodal domain theorem for graphs and partial proofs were obtained independently by different authors [39, 59, 75, 147, 169], beginning with the work of Fiedler who proved the following two results that are corollaries of the nodal domain theorem.

**Corollary 3.2 ([67]).** *The eigenfunction $f$ affording to the smallest nonzero eigenvalue of any connected graph has $\mathfrak{W}(f) = 2$ weak nodal domains.*

**Corollary 3.3 ([67]).** *The eigenfunction $f_k$ affording $\lambda_k$ has at most $k - 1$ positive weak nodal domains for $k > 1$. Consequently, $\mathfrak{W}(f) \leq 2(k - 1)$.*

Powers [147] already stated and proved the nodal domain theorem for the special case of the adjacency matrix of a graph.

The assumption that $G$ is connected is essential for the weak version of Thm. 3.1, i.e., $\mathfrak{W}(f_k) < k$ does not hold in general for disconnected graph. A counterexample is shown in Fig. 3.3. It is possible to formulate the discrete nodal domain theorem also for not necessarily connected graphs.

**Corollary 3.4.** *Let $\mathbf{M}$ be a generalized Laplacian of a graph $G$ with $c$ connected components. Then an eigenfunction $f_k$ corresponding to eigenvalue $\lambda_k$ with multiplicity $r$ satisfies*

$$\mathfrak{W}(f_k) \leq k + c - 1 \qquad and \qquad \mathfrak{S}(f_k) \leq k + r - 1 \,.$$

Notice that the number of strong nodal domains does not depend on $c$.

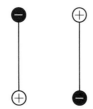

**Fig. 3.3.** Sign pattern of an eigenfunction $f_3$ of eigenvalue $\lambda_3 = 2$ (of multiplicity 2) of the disjoint union of two $K_2$: $\mathfrak{W}(f_3) = 4 > 3$.

*Proof.* Let $f_k^{(i)}$ denote the restriction of $f_k$ to the connected component $G_i$ of $G$. Either $f_k^{(i)}$ vanishes on $G_i$ or $\mathbf{M}^{(i)} f_k^{(i)} = \lambda_{k_i}^{(i)} f_k^{(i)}$, i.e., $f_k^{(i)}$ is an eigenfunction of the Laplacian $\mathbf{M}^{(i)}$ of the component $G_i$ with multiplicity $r_i$. Furthermore all these nonvanishing eigenfunctions $f_k^{(i)}$ are linearly independent. Of course $\lambda_{k_i}^{(i)} = \lambda_k$, and we choose the index $k_i$ to be the smallest one for which this equality holds, i.e., $\lambda_{k_i-1} < \lambda_k$ or (in the trivial case) $k_i = 1$. The position $k$ of $\lambda_k$ in the spectrum of $\mathbf{M}$ therefore satisfies

$$k > \sum (k_i - 1) = \sum k_i - q$$

where the summation runs over the $q$ components of $G$ on which $f_k$ does not vanish.

Courant's nodal domain theorem holds for each component $G_i$. The number of strong nodal domains of $f_k$ therefore satisfies

$$\mathfrak{S}(f_k) = \sum \mathfrak{S}(f_{k_i}^{(i)}) \leq \sum (k_i + r_i - 1) = \sum (k_i - 1) + \sum r_i < k + r$$

and hence $\mathfrak{S}(f_k) \leq k + r - 1$. Similarly we have

$$\mathfrak{W}(f_k) = \sum \mathfrak{W}(f_{k_i}^{(i)}) \leq \sum k_i = \sum (k_i - 1) + q < k + q \leq k + c$$

and thus $\mathfrak{W}(f_k) \leq k + c - 1$.  □

## 3.2 Proof of the Nodal Domain Theorem

The proofs in this section closely follow the preprint [48] which was later published as ref. [49] in a matrix-theoretic language.

We say two different weak (strong) nodal domains $D_1$ and $D_2$ of a function are *adjacent* if there exist vertices $v_1 \in D_1$ and $v_2 \in D_2$ such that $v_1 \sim v_2$. By the definition, if two different weak (strong) nodal domains are adjacent, then they have opposite signs.

We first prove the strong nodal domain theorem. Suppose that there are $m$ strong nodal domains, which we denote by $D_1, D_2, \ldots, D_m$ and (using Courant's idea) define

$$g_i(x) := \begin{cases} f_k(x) & \text{if } x \in D_i, \\ 0 & \text{otherwise,} \end{cases} \tag{3.2}$$

for $i = 1, \ldots, m$. None of these functions $g_i$ is identically zero. Since they have disjoint support, their linear span has dimension $m$. It follows that there exist constants $a_i \in \mathbb{R}$ such that $g := \sum_{i=1}^{m} a_i g_i$ is nonzero and satisfies $\langle g, f_j \rangle = 0$ for $j = 1, \ldots, m-1$. Without loss of generality we can assume $\langle g, g \rangle = 1$. Cor. 2.5 implies $\langle g, \mathbf{M}g \rangle \geq \lambda_m$.

Let us now introduce a function $a \colon V \to \mathbb{R}$ defined by $a(x) = a_i$ if $x \in D_i$ and $f_k(x) \neq 0$, and $a(x) = 0$ otherwise. By construction, $g(x) = a(x)f_k(x)$ for all $x \in V$. Starting from (2.5) we compute

$$g(x)(\mathbf{M}g)(x) = a^2(x)f_k(x) \left[ \sum_{y \sim x} (-M_{xy})(f_k(x) - f_k(y)) + p(x)f_k(x) \right]$$
$$+ a(x)f_k(x) \left( \sum_{y \sim x} (-M_{xy})(a(x) - a(y))f_k(y) \right) .$$

Using (2.5) again we observe that the term in the square brackets is $(\mathbf{M}f_k)(x)$ which is equal to $\lambda_k f_k(x)$. Summing over $x$ and symmetrizing the second term we therefore obtain

$$\langle g, \mathbf{M}g \rangle = \lambda_k \langle g, g \rangle + \sum_{x \in V} \sum_{y \sim x} (-M_{xy})f_k(x)f_k(y)a(x)(a(x) - a(y))$$
$$= \lambda_k + \frac{1}{2} \sum_{xy \in E} (-M_{xy})f_k(x)f_k(y)(a(x) - a(y))^2 . \tag{3.3}$$

A term in the sum vanishes if $f_k(x) = 0$ or $f_k(y) = 0$. If $f_k(x)f_k(y) > 0$ and $x \sim y$, i.e. $M_{xy} < 0$, then $x$ and $y$ lie in the same nodal domain and thus $a(x) = a(y)$, i.e., the corresponding contribution to the sum vanishes as well. The only remaining terms are those for which $f_k(x)f_k(y) < 0$ (and $M_{xy} < 0$). Thus the sum cannot be positive and we have $\langle g, \mathbf{M}g \rangle \leq \lambda_k$. Recalling $\lambda_m \leq \langle g, \mathbf{M}g \rangle$ from the previous paragraph we conclude $\lambda_m \leq \lambda_k$. Since $\lambda_k < \lambda_{k+r}$ we have $\lambda_m < \lambda_{k+r}$, and $m < k+r$, i.e., $\mathfrak{S}(f_k) \leq k+r-1$.

In order to prove the weak nodal domain theorem we assume that there are $m$ weak nodal domains, which we again denote by $D_1, D_2, \ldots, D_m$. As above, we define functions $g_i(x)$, $g(x) = \sum_{i=1}^{m} a_i g_i(x)$ and $a(x)$, and show $\lambda_m \leq \langle g, \mathbf{M}g \rangle \leq \lambda_k$. Now suppose that $m > k$. Then $\lambda_k \leq \lambda_m$, hence $\lambda_m \leq \langle g, \mathbf{M}g \rangle \leq \lambda_k \leq \lambda_m$ and consequently $\lambda_m = \lambda_k$. Thus by Cor. 2.5, $g$ must be an eigenfunction of eigenvalue $\lambda_k$. To finish the proof we need a continuation result for the coefficients $a_i$, which could be seen as a discrete analog of the unique continuation principle for eigenfunctions of PDEs.

**Lemma 3.5.** *Suppose $m \geq k$, and two of the weak nodal domains $D_i$ and $D_j$ of $f_k$ are adjacent (and thus have different sign). If $a_i \neq 0$ then $a_i = a_j$.*

*Proof.* Let $x \in D_i$ and $y \in D_j \setminus D_i$ be two adjacent vertices. If $f_k(x) \neq 0$ then $f_k(y) \neq 0$ since otherwise $y \in D_i \cap D_j$. Thus $f_k(x)f_k(y) < 0$ and from $\lambda_m = \lambda_k$ it follows that the sum in (3.3) must be 0 and we conclude $a(x) = a(y)$. Therefore $a_i = a(x) = a(y) = a_j$.

Now assume $f_k(x) = 0$ and define $h(v) = f_k(v) - \frac{1}{a_i}g(v)$. Both $f_k$ and $g$ are eigenfunctions of $\lambda_k$ and both vanish on $x$. Using (2.5) we find

$$0 = \lambda_k h(x) = (\mathbf{M}h)(x) = \sum_{v \sim x}(-M_{xv})(h(x) - h(v)) + p(x)h(x)$$

$$= \sum_{v \in B} M_{xv}h(v) = (1 - \tfrac{a_j}{a_i})\sum_{v \in B} M_{xv}f_k(v) \,,$$

where $B = \{v \sim x \colon v \notin D_i\}$ is the set of all neighbors of $x$ that belong to weak nodal domains different from $D_i$, i.e., $f_k(v)$ has the same strict sign on $B$. Since $B \cap D_j$ contains at least the vertex $y$ we see that the last sum above cannot vanish. Thus the prefactor $1 - a_j/a_i$ must vanish, i.e. $a_i = a_j$.    □

Since $g$ is an eigenfunction of $\mathbf{M}$ at least one coefficient, say $a_1$ must be nonzero. By the above lemma the coefficients $a_i$ for all weak nodal domains that are adjacent to $D_1$ must be equal to $a_1$. Since $G$ is connected we can conclude by repeating this argument that $a_i = a_1$ for $i = 1, \dots, m$ and hence, $g = a_1 \sum_{i=1}^{m} g_i = a_1 f_k$. But $g$ is orthogonal to $f_j$ for $j = 1, \dots, m-1$ by construction; if $m > k$, $g$ is therefore orthogonal to $f_k$, a contradicting to $g = a_1 f_k$. Thus $m \leq k$, which implies $\mathfrak{W}(f_k) \leq k$.    □

*Remark 3.6.* The published literature contains a number of incomplete or even incorrect statements and proofs of nodal domain theorems. A reasons for this peculiar history appears to be – at least in part – the "intuitively obvious" but false conjecture that $\mathfrak{S}(f_k)$ should be at most $k$.

Colin de Verdière [39] stated that any eigenfunction corresponding to $\lambda_k$ has at most $k$ weak nodal domains; his proof however relies on assertions that are stated without proof. In a later survey Colin de Verdière [41] did not mention the theorem. Friedman's [75] proof of the weak nodal domain theorem is also incomplete.

There are several attempts to improve the bound of the strong nodal domain theorem for special cases (see e.g. Sect. 3.6). However, not all are correct. Theorem 2.4 in [75] and Thm. 4.4 in [169] can be rephrased in the following way: *If an eigenfunction $f_k$ corresponding to $\lambda_k$ has more than $k$ strong sign graphs, then there is no pair of adjacent vertices $u$ and $v$ that join strong nodal domains.* Figure 3.4 shows a counterexample to this claim: the eigenfunction shown has $6 > k = 5$ strong nodal domains.

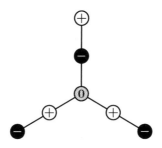

**Fig. 3.4.** Sign pattern of a particular eigenfunction $f$ corresponding to $\lambda_5$ (multiplicity $r = 2$) of the Laplacian of $G$ with $\mathfrak{S}(f) = 6 > k = 5$.

Duval and Reiner [59] attempted to show that an eigenfunction corresponding to $\lambda_k$ has no more than $k$ strong sign graphs, despite the earlier counterexample given in [75], see Fig. 3.1. Li and Guan [127] claimed that a generalization of (3.3) on which other (valid) theorems in [59] are based is incorrect and give an alternative proof. However, these authors also claim that their method can be used to prove the incorrect upper bound $k$ for strong nodal domains.

Friedman and Tillich [76] conjecture that *many concepts in analysis translate almost immediately to* their setting (see Sect. 2.4). The necessity to distinguish between strong and weak nodal domains hints at subtle differences in the results, however. A more spectacular example for the differences between the continuous and the discrete world will be discussed in detail in Chap. 6.

## 3.3 Algebraic Connectivity, Fiedler Vectors and Perron Branches

Theorem 2.3 implies that the second smallest eigenvalue $\lambda_2(G)$ of the Laplacian $\mathbf{L}(G)$ of a graph $G$ is 0 if and only if $G$ is disconnected. Thus Fiedler [65] called eigenvalue $\lambda_2(G)$ the *algebraic connectivity* of $G$. He showed that $\lambda_2(G)$ is closely related to $\mathbf{v}(G)$ and $\mathbf{e}(G)$, the vertex and edge connectivities of $G$, respectively.

**Lemma 3.7 ([65]).** *Let $G$ be a graph, let $G_k$ arise from $G$ by removing $k$ vertices from $G$ and all incident edges, then*

$$\lambda_2(G_k) \geq \lambda_2(G) - k .$$

*Proof.* First let $k = 1$ and remove vertex $v$. Define the graph $\hat{G}$ by inserting into $G$ edges from $v$ to all other vertices. By Cor. 2.5, $\lambda_2(\hat{G}) \geq \lambda_2(G)$. Moreover

$$\mathbf{L}(\hat{G}) = \begin{pmatrix} \mathbf{L}(G_1) + \mathbf{I} & -\mathbf{1} \\ -\mathbf{1}^\mathsf{T} & n-1 \end{pmatrix}$$

where $\mathbf{1} = (1, \ldots, 1)^\mathsf{T}$ with $n-1$ ones. If $f$ is an eigenfunction of $\lambda_2(G_1)$ then

$$\mathbf{L}(\hat{G})\hat{f} = (\lambda_2(G_1) + 1)\hat{f}$$

for $\hat{f} = \begin{pmatrix} f \\ 0 \end{pmatrix}$, i.e., $\lambda_2(G_1) + 1$ is a strictly positive eigenvalue of $\mathbf{L}(\hat{G})$. Thus we find

$$\lambda_2(G) \leq \lambda_2(\hat{G}) \leq \lambda_2(G_1) + 1$$

so that the inequality is fulfilled for $k = 1$. The general case now follows by induction with respect to $k$.                    □

**Theorem 3.8 ([65]).** *Let $G$ be a graph that is not complete. Then*

$$\lambda_2(G) \leq \mathbf{v}(G) \leq \mathbf{e}(G) .$$

*Proof.* Let $V_1$ be a vertex cut of graph $G(V, E)$ with $\mathbf{v}(G)$ vertices. Since $G$ is not complete, the graph $G_1$ obtained from $G$ by removing the vertices from $V_1$ and all incident edges is nonvoid and not connected. By Lemma 3.7 $0 = \lambda_2(G_1) \geq \lambda_2(G) - \mathbf{v}(G)$, and thus the first inequality follows. The second inequality follows from a deeper result of Whitney [176]: For any pair $w$, $w'$ of vertices of $G$ there exist $\mathbf{v}(G)$ paths between $w$ and $w'$ in $G$, no two of them having any vertices in common (except their endpoints).                    □

*Remark 3.9.* Notice that $\lambda_2(K_n) = n$ for the complete graph $K_n$ with $n$ vertices. However, for convenience Fiedler [65] has set $\mathbf{v}(K_n) = n - 1$.

*Remark 3.10.* The bound in Thm. 3.8 can be improved. Kirkland gives a sharp upper bound for the algebraic connectivity in terms of the number $n$ of vertices and the number of cutpoints of the graph, see [112, 116].

There is also a lower bound for the algebraic connectivity. The proof uses a result for symmetric doubly stochastic matrices, see [64].

**Theorem 3.11 ([65]).** *For any graph $G$ with $n$ vertices*

$$\lambda_2(G) \geq 2 \left(1 - \cos \frac{\pi}{n}\right) \mathbf{e}(G) .$$

*Equality holds if and only if $G$ is a path.*

Theorems 3.8 and 3.11 can be generalized for Laplacians on weighted graphs, see [68].

An eigenfunction affording the second smallest eigenvalue of a graph Laplacian is called a *Fiedler vector* of G. As mentioned in the introduction, Fiedler vectors have many applications in various fields of application, where they form the basis of heuristic approaches to solve hard combinatorial optimization problems including graph bi-partitioning [146] and spectral clustering [15]. Therefore they have been extensively investigated.

Fiedler [67] has called eigenfunctions $f_2$ corresponding to $\lambda_2$ the *characteristic valuations* of G. If G is connected then there is exactly one positive and one negative nodal domain (Cor. 3.2) which are separated by *characteristic vertices* u where the eigenfunction $f_2$ vanishes, i.e., $f_2(u) = 0$ and *characteristic edges* uv where $f_2$ changes sign, i.e., $f_2(u)f_2(v) < 0$.

**Theorem 3.12 ([67, Thm. (3,14)] and [114, Thm. 6]).** *Let G be a tree and f one of its Fiedler vectors. Then two cases can occurs:*

*(1) If $f(v) \neq 0$ for all $v \in V$ then there exists exactly one edge uv (the characteristic edge) such that $f(u) > 0$ and $f(v) < 0$. Moreover, f is increasing and concave on every path that starts in u and does not contain v, and decreases and convex on every path starting in v and not containing u.*

*(2) Otherwise, let $V_0 = \{v \in V : f(v) = 0\}$. Then the graph induced by $V_0$ is connected and there is exactly one vertex $u \in V_0$ (the characteristic vertex) that is adjacent to a vertex that does not belong to $V_0$. Moreover, f is either increasing and concave, decreasing and convex, or identically zero on any path starting from u.*

*That is, a tree possesses only one characteristic element which is either a vertex or an edge.*

Merris [132] showed that the occurrence of case (1) or (2) does not depend on the Fiedler vector. He classified trees that contain a characteristic vertex as *Type I*, and those that contain a characteristic edge as *Type II*.

Algebraic connectivity, Fiedler vectors and the set of characteristic vertices and edges has been investigated by various authors, see e.g. [7–9, 113–115, 132, 140]. Merris [132] has described the *characteristic set* of trees. For example he showed that if v is a characteristic vertex of some Fiedler vector f of a tree G then it is a characteristic vertex of every Fiedler vector of this tree; see Sect. 4.1.1 for an extension of this result for arbitrary eigenvalues of G. Bapat and Pati [8] have given a more general result and showed that a statement similar to Thm. 3.12 holds for the special case of unicyclic graphs (i.e., connected graphs that have exactly one cycle).

It is remarkable that the cardinality of the characteristic set can be sharply bounded by means of the cyclomatic number of the graph (i.e., the dimension of the cycle space, see [30]) that is given by $|E| - |V| + 1$. The bound is sharp for arbitrarily large numbers of vertices.

**Theorem 3.13 ([8]).** *Let $C$ be the characteristic set of a Fiedler vector of a connected graph $G$. Then $1 \le |C| \le |E| - |V| + 2$. In particular if $G$ is a tree then $|C| = 1$.*

The main concept for investigating the structure of Fiedler vectors is the notion of *Perron components* first introduced by Kirkland et al. [114] for trees (*Perron branches*).

We first describe the closely related notion of *geometric nodal domains* that is based on the concept of calculus on graphs (Sect. 2.4). Let $W$ be the characteristic set of some Fiedler vector on a graph $G$ that entirely consists of vertices. Then each component $D_0$ of $G \setminus W$ is a strong sign graph (nodal domain) of $f$. The principal submatrix of $\mathbf{L}(G)$ corresponding to this component is simply the Dirichlet matrix $\mathbf{L}^\circ(D)$ of $D$, see Sect. 1.5, where $D$ is the graph with boundary that consists of $D_0$ (interior vertices and edges) and adjacent vertices from $W$ as its boundary vertices and the corresponding edges as its boundary edges. If $W$ contains a characteristic edge of a Fiedler vector $f$ we proceed as follows. Let $\mathcal{G}$ be the geometric realization of graph $G$. Eigenfunction $f$ can be extended to an eigenfunction $\bar{f}$ of the "continuous Laplacian" $\mathcal{L}(\mathcal{G})$ by a linear interpolation of $f$ on each edge, see Sect. 2.4. A *geometric nodal domain* $\mathcal{D} \subseteq \mathcal{G}$ is then the closure of a component of $\mathcal{G} \setminus \{x \colon \bar{f}(x) = 0\}$. Furthermore, there exists a graph $D$ with boundary whose geometric realization is exactly $\mathcal{D}$ and whose boundary vertices are the points on which $\bar{f}$ vanishes. In abuse of language we call the graph $D$ a geometric nodal domain of $f$. Notice that there is a natural homomorphism from $D$ into $G$ where boundary edges are mapped to characteristic edges or edges that are incident to characteristic vertices. A boundary edge of a geometric nodal domain may have weights. Let $vu$ the corresponding characteristic edge in $G$ with $v \in D_0$ and $u \notin D_0$ then the boundary edge $vu'$ in $D$ has length $f(v)/(f(v) - f(u))$ and thus weight $w_{vu'} = (f(v) - f(u))/f(v)$. It is obvious that geometric nodal domains can be defined analogously for arbitrary eigenfunctions.

**Lemma 3.14.** *If $f$ is an eigenfunction corresponding to some eigenvalue $\lambda$ of $\mathbf{L}(G)$, and let $D$ be a geometric nodal domain of $f$. Then $\lambda^\circ(D) = \lambda$ with $f$ restricted to the interior vertices of $D$ as its eigenfunction, where $\lambda^\circ(D)$ denotes the first (Dirichlet) eigenvalue of $\mathbf{L}^\circ(D)$.*

*Proof.* Let $V_0 \cup \partial V$ be the set of interior and boundary vertices of $D$ and $E_0 \cup \partial E$ the set of interior and boundary edges. The Dirichlet matrix $\mathbf{L}^\circ$ on $D$ is the principal submatrix corresponding to $V_0$ of the Laplacian $\mathbf{L}_w$ on the weighted graph $D$. By (2.8) for every $v \in V_0$,

$$(\mathbf{L}_w f)(v) = \sum_{u \sim v} w_{vu}(f(v) - f(u)) = \sum_{V_0 \ni u \sim v} (f(v) - f(u)) + \sum_{\partial V \ni u' \sim v} w_{vu'} f(v)$$

$$= \sum_{u \sim v} (f(v) - f(u)) = (\mathbf{L}f)(v) = \lambda_2(G)\, f(v).$$

Thus $f$ restricted to $D$ is an eigenfunction of $\mathbf{L}^\circ(D)$ for eigenvalue $\lambda_2(G)$. Since $f$ does not change sign, $\lambda_2(G)$ must be the lowest eigenvalue of $\mathbf{L}^\circ(D)$.    □

An immediate corollary of this lemma (together with Prop. 2.7) is that the lowest eigenvalue of a principal submatrix that corresponds to a particular strong sign graph $S$ is equal or greater than the corresponding eigenvalue $\lambda$ of $\mathbf{L}(G)$. Equality holds if and only if the characteristic set entirely consists of vertices. If there are some characteristic edges incident to $S$ then this inequality is strict. Moreover, when we add all characteristic edges (and its end vertices) that are incident to $S$, the lowest eigenvalue of the corresponding principal submatrix is strictly smaller than $\lambda$. For the algebraic connectivity of a graph a converse result also holds.

**Proposition 3.15 ([8]).** *Let $W$ be any set of vertices of a graph $G$ such that $G \setminus W$ is non-empty and not connected. Let $G_1$ and $G_2$ be two components of $G \setminus W$ and $\mathbf{L}_1$ and $\mathbf{L}_2$ be the principal submatrices of $\mathbf{L}(G)$ corresponding to $G_1$ and $G_2$, respectively. Suppose $\lambda_1(\mathbf{L}_1) \leq \lambda_1(\mathbf{L}_2)$ then the following holds: Either $\lambda_1(\mathbf{L}_2) > \lambda_2(G)$ or $\lambda_1(\mathbf{L}_1) = \lambda_1(\mathbf{L}_2) = \lambda_2(G)$. Moreover, if $W$ consists of characteristic vertices only, then the latter condition is always satisfied.*

Following Bapat and Pati [8] we call a component of $G \setminus W$ of a proper subset $W \subset G$ a *Perron component* if the lowest eigenvalue of its corresponding principle submatrix of $\mathbf{L}(G)$ does not exceed $\lambda_2(G)$.

Obviously every graph has at least two Perron components by Cor. 3.2. In addition to Thm 3.1 if a graph $G$ has $t$ ($\geq 2$) Perron components then the multiplicity of $\lambda_2(G)$ is at least $t - 1$, see Bapat et al. [7]. However, the cycle $C_4$ of length 4 shows that this bound is not sharp. If the characteristic set contains only vertices then one can obtain $t - 1$ linearly independent Fiedler vectors by means of $f_2$ restricted to each of the components [7].

**Theorem 3.16 ([9, Thm. 10]).** *Let $G$ be a connected graph and $f_2$ a Fiedler vector with characteristic set $W$ consisting of vertices only. Suppose there are $t$ Perron components of $G$ at $W$. Then the following is equivalent:*

*(i) The multiplicity of $\lambda_2(G)$ is $t - 1$.*
*(ii) For each Fiedler vector the characteristic set is $W$.*
*(iii) For each Fiedler vector the characteristic set consists of vertices only.*

It is easy to construct an arbitrary type I tree. Take two (or more) copies of a tree with root $v_0$ and identify the roots in each copy. The resulting tree has a Fiedler vector with $v_0$ as characteristic vertex and the Perron branches are all isomorphic. Kirkland [111] showed how type I trees can be constructed with non-isomorphic Perron branches.

Finally, we notice that many of the above results hold analogously for generalized Laplacians.

## 3.4 Some Results for Multiple Eigenvalues

We have seen in Sect. 3.1 that the upper bound $\mathfrak{S}(f_k) \leq k + r - 1$ in Thm. 3.1 cannot be improved in general. However, there always exists an eigenfunction $f_k$ with at most $k$ strong nodal domains.

**Lemma 3.17.** $\mathbf{M}(G)$ *has an eigenfunction* $f_k$ *corresponding to eigenvalue* $\lambda_k$ *with* $\mathfrak{S}(f_k) \leq k$ *for all* $k$.

*Proof.* Suppose $f_k$ has $\mathfrak{S}(f_k) = m > k$. Following Sect. 3.2 we can construct a nonzero function $g(x) = \sum_{i=1}^{k} a_i\, g_i(x)$ satisfying $\langle g, f_j \rangle = 0$ for $j = 1, \ldots, k-1$ and $\langle g, g \rangle = 1$, where the functions $g_i$ are as defined in (3.2). By construction we have $\mathfrak{S}(g) \leq k$. By the same argument we see $\langle g, \mathbf{M} g \rangle = \lambda_k$ and thus Cor. 2.5 implies that $g$ is an eigenfunction of $\mathbf{M}(G)$ to $\lambda_k$.    □

Gladwell and Zhu [81] proved stronger results.

**Theorem 3.18 ([81]).** *There exist orthogonal eigenfunctions* $f_k$ *of* $\mathbf{M}(G)$ *such that* $\mathfrak{S}(f_k) \leq k$ *for* $k = 1, \ldots, n$.

*Proof.* By Thm. 3.1 we only have to deal with multiple eigenvalues. We proceed by induction. Assume that we have eigenfunctions $f_k, \ldots, f_{k+j-1}$, $1 \leq j < r$, with the desired property. Then there is some eigenfunction $h_{k+j}$ of $\lambda_k$ which is orthogonal to $f_k, \ldots, f_{k+j-1}$. If $\mathfrak{S}(h_{k+j}) \leq k + j$ we set $f_{k+j} = h_{k+j}$. Otherwise we construct such an eigenfunction $f_{k+j}$ as in the proof of Lemma 3.17.    □

**Theorem 3.19 ([81]).** *Suppose that* $\lambda_k$ *is eigenvalue with multiplicity* $r$ *and eigenspace* $\mathcal{E}_k$. *Then there exists a basis* $\{f_k, \ldots, f_{k+r-1}\}$ *of* $\mathcal{E}_k$ *such that* $\mathfrak{S}(f_j) \leq k$ *for all* $j = k, \ldots, k + r - 1$.

*Proof.* We proceed as in Thm. 3.18. Again we have an eigenfunction $h_{k+j}$ of $\lambda_k$ which is now linearly independent of $f_k, \ldots, f_{k+j-1}$. If $\mathfrak{S}(h_{k+j}) \leq k$ we set $f_{k+j} = h_{k+j}$. Otherwise, $\mathfrak{S}(h_{k+j}) = m > k$ and we define the functions $g_i$ as $h_{k+j}$ restricted to the $i$-th strong nodal domain of $h_{k+j}$. We construct the set of all linear combinations $g = \sum_{i=1}^{m} a_i\, g_i$ which satisfy $\langle g, f_s \rangle = 0$ for all $s = 1, \ldots, k - 1$. As these are $k - 1$ constraints this set can be spanned by linear combinations where at most $k$ coefficients $a_i$ are nonzero and thus each of these has at most $k$ strong nodal domains. Moreover, at least one of these functions must be linearly independent from $f_k, \ldots, f_{k+j-1}$, as this set contains $h_{k+j}$.    □

When examine the proof of the nodal domain theorem in Sect. 3.2 then it follows from (3.3) that $a(x) = a(y)$ whenever $x$ and $y$ are adjacent and belong to different strong nodal domains. Thus we have the following result.

**Theorem 3.20 ([81]).** *If an eigenfunction* $f_k$ *affording eigenvalue* $\lambda_k$ *has* $k + s$ *strong nodal domains, where* $s \geq 1$, *then* $G \setminus \{v \colon f_k(v) = 0\}$ *consists of at least* $s + 1$ *connected components.*

*Proof.* Following Sect. 3.2 we can construct a nonzero eigenfunction $g$ affording $\lambda_k$ which vanishes on (at least) $s$ of the strong nodal domains of $f_k$. We can choose each of these nodal domains in different components of $G \setminus \{v \colon f_k(v) = 0\}$. By the above argument $g$ must vanish on each of these components. Thus there must be at least $s + 1$ such components since otherwise $g$ would be identically zero.    □

This theorem has some interesting implications.

**Corollary 3.21.** *Let $f_k$ be an eigenfunction of $\mathbf{M}(G)$ with eigenvalue $\lambda_k$.*

*(i) If $G$ is connected and $f_k$ does not vanish on any vertex, then $\mathfrak{S}(f_k) \leq k$.*

*(ii) Each of the connected components of $G \setminus \{v \colon f_k(v) = 0\}$ has at most $k$ strong nodal domains.*

*(iii) If one of these components has $k$ strong nodal domains, then $\mathfrak{S}(f_k) \geq k + 2$.*

## 3.5 The Courant-Herrmann Conjecture

A footnote on p. 454 in [44] claims that the nodal domain theorem for manifolds (p. 29) can be generalized in the following way: *Any linear combination of the first $n$ eigenfunctions divides the domain, by means of its nodes, into no more than $n$ subdomains.* For a proof the reader is referred to Herrmann's 1932 dissertation in Göttingen. We will refer to this statement as the *Courant-Herrmann conjecture* (CHC). Gladwell and Zhu [82] reported, however, that neither in this work nor in any of Herrmann's subsequent publications such a result is stated, let alone proved. Indeed, they showed that the statement is false in general, and gave a simple counterexample.

For graphs we naturally have two versions of the CHC. Let $f = \sum_{i=1}^{k} a_i f_i$ be some linear combination of the eigenfunctions $f_1, \ldots, f_k$ of a (generalized) Laplacian, and $r$ the multiplicity of $\lambda_k$. Then

- *weak CHC:*   $\mathfrak{W}(f) \leq k$,
- *strong CHC:*   $\mathfrak{S}(f) \leq k + r - 1$.

Both statements do not hold in general. For example, let $G$ be the star with $n \geq 5$ vertices of Fig. 3.1. Then $f_1 = \mathbf{1}$ is an eigenfunction corresponding to eigenvalue $\lambda_1 = 0$ of the Laplacian $\mathbf{L}(G)$ and there exists an eigenfunction $f_2$ affording eigenvalue $\lambda_2$ with one positive vertex, $n - 2$ negative vertices, and the zero vertex in the center of $G$. Obviously $f = \varepsilon f_1 + f_2$ for sufficiently small $\varepsilon > 0$ has one weak positive nodal domain and $n - 2$ weak negative nodal domains, i.e., $\mathfrak{W}(f) = \mathfrak{S}(f) = n - 1 > k = 2$, a contradiction to the weak CHC.

Gladwell and Zhu [82] give an example with a particular generalized Laplacian for the star graph which does not have multiple eigenvalues. Their example also yields $\mathfrak{W}(f) = \mathfrak{S}(f) = n - 1 > k + r - 1 = 2$.

**Problem 3.22.** For which graphs does the (weak or strong) Courant-Herrmann conjecture hold?

## 3.6 Improvements for Special Cases

Neither the weak nor the strong version of the nodal domain theorem can be strengthened without additional assumptions, because the discrete nodal domain theorem is sharp for paths.

**Theorem 3.23 (Gantmacher and Krein [78]).** *The eigenvalues of a generalized Laplacian* **M** *of a path are all simple, and the eigenfunction $f_k$ corresponding to eigenvalue $\lambda_k$ has exactly $k$ (strong and weak) nodal domains, i.e., $\mathfrak{W}(f_k) = \mathfrak{S}(f_k) = k$.*

An example where $f_k$ has more than $k$ strong nodal domain is e.g. given by Friedman [75]: a star on $n$ vertices has a second eigenfunction with $n - 1$ strong nodal domain, which is exactly the upper bound $k + r - 1$; see Fig. 3.1.

However, as for manifolds, the nodal domain theorem for graphs does not provide a sharp inequality for all graphs. For manifolds equality for every eigenvalue holds only in dimension one, i.e. for a string. For spheres with the standard metric a sharp lower bound on the number of nodal domains exists [123] but so far no sharp upper bounds are available, see e.g. [3, 107, 108, 124]. For graphs the situation is similar. Improved general upper bounds for $\mathfrak{W}(f_k)$ and $\mathfrak{S}(f_k)$ for trees, cographs, and hypercubes will be discussed in the next chapter. In the remainder of this section we consider a few results for the eigenfunctions of particular eigenvalues.

Let us first consider the largest eigenvalue of a connected bipartite graph $G$.

**Theorem 3.24 (Roth [152]).** *Let $G(V_1 \cup V_2, E)$ be a connected bipartite graph with $n = |V_1 \cup V_2|$ vertices and let* **M** *be a generalized Laplacian of $G$. Then there is an eigenfunction $f$ corresponding to the largest eigenvalue of* **M**, *such that $f$ is positive on $V_1$ and negative on $V_2$ or vice versa and hence satisfies $\mathfrak{W}(x) = \mathfrak{S}(x) = n$.*

*Proof.* The largest eigenvalue $\lambda_n$ of **M** is determined by the maximum of the Rayleigh quotient $\mathcal{R}_\mathbf{M}(f)$. We may assume that $f$ is normalized so that $\langle f, f \rangle = 1$; thus by (2.7) we have

$$\mathcal{R}_\mathbf{M}(f) = \sum_{x \in V} M_{xx} f(x)^2 + 2 \sum_{xy \in E} M_{xy} f(x) f(y) \ .$$

Let $f_n$ be an eigenfunction affording $\lambda_n$ and define $g(x) = |f_n(x)|$ if $x \in V_1$ and $g(x) = -|f_n(x)|$ if $x \in V_2$. We have $\mathcal{R}_\mathbf{M}(g) \geq \mathcal{R}_\mathbf{M}(f_n)$; this inequality is strict if and only if there is an edge $xy \in E$ such that $f_n(x)f_n(y) > 0$. Since $f_n$ maximizes $\mathcal{R}_\mathbf{M}$ we have $f_n(x)f_n(y) \leq 0$ for all $xy \in E$. Therefore $g$ is an eigenfunction of $\lambda_n$.

Now suppose $g(x) = 0$ for some $x \in V_1$. Then $\sum_{y \sim x} M_{xy} g(y) = \lambda_n g(x) = 0$. Since all neighbors of $x$ are contained in $V_2$ this implies $g(y) = 0$ for all $y \sim x$. Repeating the argument shows that $g$ must vanish. Thus $g(x) > 0$ and hence

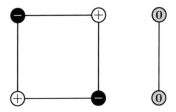

**Fig. 3.5.** Sign pattern of an eigenfunction corresponding to the largest eigenvalue of the disconnected bipartite graph $C_4 \cup K_2$: $\mathfrak{S}(f_6) = 4$.

either $f_n = g$ or $f_n = -g$. Since any two neighboring vertices have opposite strict signs we see that each vertex $x \in V$ is a strong nodal domain, and the theorem follows. □

The discrete nodal domain theorem implies the following

**Corollary 3.25.** *The largest eigenvalue of a generalized Laplacian of a connected bipartite graph is simple.*

*Proof.* By Thm. 3.1 any eigenfunction corresponding to eigenvalue $\lambda_{n-1}$ has at most $n-1$ weak nodal domains. If the largest eigenvalue is not simple, i.e., $\lambda_{n-1} = \lambda_n$, then the eigenfunction of Thm. 3.24 would also be an eigenfunction corresponding to $\lambda_{n-1}$, a contradiction. □

*Remark 3.26.* Again we stress the importance of connectedness here. Theorem 3.24 does not hold if the graph $G$ is disconnected, see Fig. 3.5 for a counterexample. Moreover, suppose $G$ has at least 2 connected components and there exists such an eigenfunction $f$ with $n$ (strong) nodal domains. Then we can restrict $f$ to each of the connected components. Obviously these are linearly independent and each of these is an eigenfunction. Thus the largest eigenvalue cannot be simple.

It is quite obvious that a graph $G$ cannot have $n$ nodal domains if it is not bipartite. It fact one easily finds the following results.

**Theorem 3.27 ([22]).** *Let $G(V, E)$ be a connected graph and $H$ be an induced bipartite subgraph of $G$ with maximum number of vertices. Then for any eigenfunction $f$ of a generalized Laplacian $\mathbf{M}(G)$, $\mathfrak{S}(f) \leq |V(H)|$.*

*Proof.* We delete all zero vertices and for each strong nodal domain we delete all but one vertex. The subgraph induced by the remaining vertices is bipartite and the result follows, since $H$ is a bipartite induced subgraph with maximum number of vertices. □

Unfortunately, to find such an induced bipartite graph of $G$ is a well known NP-complete problem (see, e.g., [80]). In general the upper bound of Thm. 3.27 is not sharp for the graph Laplacian $\mathbf{L}$, see Fig. 3.6 for a counterexample. However, we can show that it is sharp for the generalized Laplacians of every given graph.

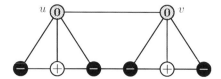

**Fig. 3.6.** Sign pattern of an eigenfunction with maximal number of strong nodal domains: $\mathfrak{S}(f) = 5 < 6 = |V(H)|$. One easily checks that for all simple eigenvalues there are at most 4 strong nodal domains. For the only multiple eigenvalue $\lambda_5 = 4$ (multiplicity $r = 2$) we have $f(u) = f(v)$. If both are nonzero $\mathfrak{S}(f) \leq 4$; otherwise we have 5 strong nodal domains.

**Theorem 3.28 ([22]).** *Let $G$ be a connected graph and $H$ be a maximal induced bipartite subgraph of $G$, then there exists a generalized Laplacian $\mathbf{M}(G)$ such that $\mathbf{M}(G)$ has an eigenfunction $f$ with $|V(H)|$ strong nodal domains.*

*Proof.* Let $H$ be a maximum induced bipartite subgraph of $G$ with components $C_1, \ldots, C_k$ and let $R$ be the set of remaining vertices of $G$. Let $\mathbf{M}_1, \ldots, \mathbf{M}_k$ be generalized Laplacians of $C_1, \ldots, C_k$ such that diagonal elements of $\mathbf{M}_i$ are positive. By Theorem 3.24 the largest eigenvalue $\mu_i$ of $\mathbf{M}_i$ has an eigenfunction $f_i$ with $\mathfrak{S}(f_i) = |V(C_i)|$. The eigenvalues $\mu_i$ are positive, since $\mathrm{tr}(\mathbf{M}_i) > 0$. Thus we can assume without loss of generality that $\mu_i = 1$ (otherwise replace $\mathbf{M}_i$ by $\frac{1}{\mu_i}\mathbf{M}_i$). We now define a generalized Laplacian for $G$ by

$$
\mathbf{M} = \begin{pmatrix}
\mathbf{M}_1 & 0 & \cdots & 0 & \mathbf{B}_1^{\mathsf{T}} \\
0 & \mathbf{M}_2 & \cdots & 0 & \mathbf{B}_2^{\mathsf{T}} \\
\vdots & \vdots & \ddots & \vdots & \vdots \\
0 & 0 & \cdots & \mathbf{M}_k & \mathbf{B}_k^{\mathsf{T}} \\
\mathbf{B}_1 & \mathbf{B}_2 & \cdots & \mathbf{B}_k & \mathbf{M}_R
\end{pmatrix}
\tag{3.4}
$$

where $\mathbf{M}_R$ is some generalized Laplacian on the graph induced by $R$, and the $\mathbf{B}_i$ matrices with nonpositive entries. Notice that each vertex $v \in R$ has (at least) two neighbors $w_1$ and $w_2$ in some $C_j$ such that $f_j(w_1)$ and $f_j(w_2)$ have opposite (strict) sign, since otherwise we could construct a new bipartite graph with more vertices then $H$. Thus we can choose $\mathbf{B}_1, \ldots, \mathbf{B}_k$ such that $\mathbf{B}_1 f_1(v) + \cdots + \mathbf{B}_k f_k(v) = 0$. Now construct a function $f$ by $f(v) = f_i(v)$ if $v \in C_i$ and $f(v) = 0$ if $v \in R$. Then a straightforward computation gives $(\mathbf{M}f)(v) = f_i(v) = f(v)$ if $v \in C_i$ and $(\mathbf{M}f)(v) = (\mathbf{B}_1 f_1 + \cdots + \mathbf{B}_k f_k)(v) = 0 = f(v)$ if $v \in R$. Hence $\mathbf{M}f = f$ and $f$ is an eigenfunction with $\sum_{i=1}^{k} |V(C_i)| = |V(H)|$ nodal domains. $\qquad\square$

Similarly to Thm. 3.27 there exists an upper bound for the number of weak nodal domains of an arbitrary function.

**Theorem 3.29 ([22]).** *Let $G(V, E)$ be a connected graph and $G^* = (V^*, E^*)$ be a bipartite minor with a maximum number of vertices of $G$ such that edges*

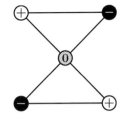

**Fig. 3.7.** Counterexample: the number of vertices of the maximum bipartite minor is not an upper bound for the number of strong nodal domains of an eigenfunction of a Laplacian; the eigenfunction $f$ has 4 strong nodal domains but $|V^*| = 3$.

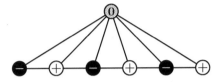

**Fig. 3.8.** Sign pattern of an eigenfunction $g$ with maximal number of strong nodal domains for a minor $H = G/uv$ of graph $G$ from Fig. 3.6: $\mathfrak{S}(g) = 6 > 5 = \mathfrak{S}(f)$.

*are only contracted in $G$ and multiple edges and loops are deleted in the resulting graph, if necessary. Then for any eigenfunction $f$ of a generalized Laplacian $\mathbf{M}(G)$, $\mathfrak{W}(f) \leq |V^*|$.*

*Proof.* We get a bipartite minor of $G$ by contracting all edges $uv$ for which $f(u), f(v) \geq 0$ and all edges $uv$ with $f(u), f(v) < 0$. Thus every weak positive nodal domain and every strong negative nodal domain of $f$ collapses into a single vertex. This minor is bipartite and the result follows, since $G^*$ is a bipartite minor with maximum number of vertices.                                      □

H. Müller [141] has remarked that finding maximal bipartite minors is also an NP-complete problem. The upper bound based on a maximal bipartite minor does not hold for strong nodal domains. For the graph in Fig. 3.7 there exists an eigenfunction of the graph Laplacian with values $(1, -1, 0, 1, -1)$. Thus it has four strong nodal domains while a maximum bipartite minor obtained by edge contractions has at most three vertices. Figure 3.8 shows an example where the maximal number of strong domains is not even monotone for minors: There exists an eigenfunction of the Laplacian of the minor $G/uv$ with 6 strong nodal domains whereas eigenfunctions of the Laplacian of the original graph $G$ in Fig. 3.6 have at most 5 strong nodal domains.

Analogously to Thm. 3.27 one could ask whether the upper bound in Thm. 3.29 is sharp. Again the graph in Fig. 3.6 serves as a counterexample for the graph Laplacian $\mathbf{L}$, as every eigenfunction has at most 5 weak nodal domains but there exists a bipartite minor $G^*$ with 6 vertices. However, it is an open question whether this bound is sharp for generalized Laplacians.

**Problem 3.30.** Let $G^*$ be a maximum bipartite minor of a graph $G$ as defined in Thm. 3.29. Is there a generalized Laplacian matrix $\mathbf{M}(G)$ such that an eigenfunction of $M(G)$ has $|V(G^*)|$ weak nodal domains?

## 3.7 Faria Vectors and Minimum Number of Nodal Domains

The number of nodal domains can be much smaller than the bound obtained from the discrete nodal domain theorem. A function $f$ is called a *Faria vector* [62], if $f$ is nonzero on only two vertices $u$ and $v$ with $f(u) = -f(v) = 1$. Obviously, Faria vectors have only two (weak or strong) nodal domains.

Two vertices $u$ and $v$ are called *twins* if every vertex $w \notin \{u, v\}$ is either adjacent to both $u$ and $v$ or to neither one of them.

**Theorem 3.31.** *A Faria vector $f$ is an eigenfunction of the Laplacian $\mathbf{L}$ of a graph $G$ if and only if $u$ and $v$ are twins in $G$. The corresponding eigenvalue is $\lambda = d(u) + 1 = d(v) + 1$ if $uv \in E(G)$ and $\lambda = d(u) = d(v)$ if $uv \notin E(G)$.*

*Proof.* It is easy to see that a Faria vector is an eigenvalue for such a graph. On the other hand if $u$ and $v$ are not twins in $G$, then there exists a zero vertex $x$ with is adjacent to exactly one nonzero vertex ($u$ or $v$), a contradiction to Lemma 2.8.                                                                                        □

There exist arbitrarily large graphs that admit Faria vectors. As an example consider graphs that have at least two vertices of degree 1 that have a common neighbor.

A type of graphs with only a small number of weak nodal domains can be immediately derived from Thm. 2.11.

**Theorem 3.32.** *If a graph $G$ with $n$ vertices has a vertex $v$ with degree $n-1$, then each eigenfunction of an eigenvalue $0 \neq \lambda < n$ of Laplacian has exactly two weak nodal domains.*

Obviously there is a trivial lower bound for the number of (strong or weak) nodal domains. The following result follows immediately from Lemma 2.8.

**Theorem 3.33.** *If $f$ is an eigenfunction of a generalized Laplacian to an eigenvalue $\lambda > \lambda_1$ then $\mathfrak{S}(f) \geq \mathfrak{W}(f) \geq 2$.*

Faria vectors show that this lower bound cannot be improved without further assumptions.

## 3.8 Sign Pattern and Nodal Domains

So far we have considered nodal domains of eigenfunctions of a given (generalized) Laplacian of a graph $G$. We could also take the opposite point of view and prescribe a sign $\{+, -, 0\}$ at each vertex of $G$. Clearly, such a *sign pattern* also induces nodal domains. One can then ask whether there is an eigenfunction of some (generalized) Laplacian $\mathbf{M}(G)$ with the same sign pattern. This is an eigenvalue problem of *sign-solvable linear systems* which are discussed in detail by Brualdi and Shader [26].

# 4

# Nodal Domain Theorems for Special Graph Classes

In Sect. 3.6 we have seen that the upper bound for the number of (strong or weak) nodal domains that is given by the discrete nodal domain theorem cannot be improved without further restrictions. On the other hand, we have seen that there exist graphs where this bound is not sharp. In general it is unknown, whether this upper bound is sharp for an arbitrary graph. The situation is similar for the (trivial) lower bound in Thm. 3.33. Furthermore, no general method is known to construct an eigenfunction of a given eigenvalue $\lambda_k$ that maximizes or minimizes the number of (strong or weak) nodal domains. In this chapter we take a closer look to the situation for trees, cographs, and product graphs (in particular to the Boolean hypercube), where it is possible to derive improved upper and lower bounds.

## 4.1 Trees

Research on the sign properties of eigenfunctions started already with the investigation of eigenfunctions of tridiagonal matrices with negative off-diagonals in the first half of the last century, see e.g. [78]. In fact, such matrices are the generalized graph Laplacians of paths and it has been shown that an eigenfunction corresponding to the $k$-th eigenvalue $\lambda_k$ has exactly $k$ (weak or strong) nodal domains, see Thm. 3.23.

Pati [143] considers the third smallest eigenvalue of a special generalized Laplacian matrix.

**Theorem 4.1 ([143]).** *Let* **M** *be a generalized Laplacian matrix of a tree* $T$, *where* $M_{vv} = \sum_{u \sim v} M_{vu}$. *If the second smallest eigenvalue is simple, then each eigenfunction belonging to the third smallest eigenvalue* $\lambda_3$ *has at most four weak nodal domains.*

In this section we characterize the maximum number of strong nodal domains of an eigenfunction corresponding to the $k$-th eigenvalue $\lambda_k$ for a tree, and present an $\mathcal{O}(n^2)$ time algorithm to find an eigenfunction with this maximum number of the strong nodal domains. In contrast, finding an eigenfunction with the minimum number of nodal domains is NP-complete.

### 4.1.1 The Maximum Number of Nodal Domains

Fiedler has shown two useful results on eigenfunctions of matrices that are related to trees[1].

**Lemma 4.2 ([66]).** *Let* $\mathbf{M}$ *be a generalized Laplacian of a tree. If* $f$ *is an eigenfunction corresponding to eigenvalue* $\lambda_k$ *which does not vanish on any vertex, then* $\lambda_k$ *is simple and there are exactly* $n - k$ *edges* $xy$ *for which* $M_{xy} f(x) f(y) < 0$.

*Proof.* We sketch the main ideas from [66] that we need to prove the result. The following two statements are usually called *Sylvester's law of inertia*: (1) For a symmetric matrix $\mathbf{B}$ and a nonsingular matrix $\mathbf{F}$, $\mathbf{B}$ and $\mathbf{F}^{\mathsf{T}}\mathbf{B}\mathbf{F}$ have the same number of positive (or negative) eigenvalues, see e.g. [100]. (2) For a given real symmetric matrix $\mathbf{B}$ let $\mathbf{C}$ be a similar matrix with quadratic form that consists of squared terms only. Then the number of positive (or negative) eigenvalues of $\mathbf{B}$ is equal to the number of positive (and negative) squared terms in the quadratic form of $\mathbf{C}$, see e.g. [78].

Now let $\mathbf{F}$ be the diagonal matrix with $F_{vv} = f(v)$, $v \in V(T)$, and let $\mathbf{B} = \mathbf{F}(\mathbf{M} - \lambda_k \mathbf{I})\mathbf{F}$. It is straightforward to see that $\mathbf{B}\mathbf{1} = 0$. Therefore the quadratic form of $\mathbf{B}$ can be computed similarly to (2.6) as $\langle g, \mathbf{B}g \rangle = \sum_{xy \in E(T)} (-B_{xy})(g(x) - g(y))^2 = \sum_{xy \in E(T)} (-M_{xy})f(x)f(y)(g(x) - g(y))^2$. Fiedler [66] showed that this quadratic form is irreducible. Hence by Sylvester's second law of inertia $\mathbf{B}$ has $n - 1$ nonzero eigenvalues and by the first law $\lambda_k$ is simple. By both laws there are exactly $n - k$ edges such that $-M_{xy}f(x)f(y) > 0$.     $\square$

As a direct consequence of the above lemma we have

**Lemma 4.3 ([66]).** *Each eigenfunction corresponding to a multiple eigenvalue of a generalized Laplacian* $\mathbf{M}$ *of a tree vanishes on at least one vertex.*

These two lemmata play an important rôle in deriving the main results of this section. We start with a special simple case where the eigenfunction does not vanish on any vertex.

**Theorem 4.4 ([19]).** *Let* $\mathbf{M}$ *be a generalized Laplacian of a tree* $T$ *on* $n$ *vertices and let* $f$ *be an eigenfunction of* $\mathbf{M}$ *with eigenvalue* $\lambda_k$ *which does not vanish on any vertex. Then* $\lambda_k$ *is simple and* $f$ *has exactly* $\mathfrak{S}(f) = k$ *strong nodal domains.*

---

[1] We remark that Fiedler [66] proved the results of Lemmata 4.2 and 4.3 for a more general matrix of a tree.

*Proof.* We divide $V(T)$ into three disjoint sets $P$, $S$, and $C$, where $P$ and $S$ denote the sets of all vertices where $f$ is positive and negative, resp., and which are incident to some edge where $f$ does not change sign. $C$ is the set of the remaining vertices. The induced subgraphs $G[P]$ and $G[S]$ are forests and consist exactly of those edges where $f$ does not change sign. Let $p$ and $s$ be the number of components of $G[P]$ and $G[S]$, respectively. Then $G[P]$ and $G[S]$ have $|P| - p$ and $|S| - s$ edges, respectively. By Lemma 4.2, $\lambda_k$ is simple and there are exactly $n - k$ edges $xy$, for which $f(x)$ and $f(y)$ have the same sign, as $M_{xy} \leq 0$. Since $\{P, S, C\}$ is a partition of $V$ we have $|P| - p + |S| - s = n - k$.

By definition $f$ must change sign on each edge between a vertex in $C$ and a vertex in $P$ or $S$. Moreover, if $f(u)$ and $f(v)$ have the same sign for $u, v \in C$, then $u$ and $v$ are not adjacent. Consequently the number of strong nodal domains of $f$ is equal to $|C| + p + s$. Thus

$$\mathfrak{S}(f) = |C| + p + s = (n - |P| - |S|) + (|P| + |S| - n + k) = k$$

as proposed.                                                                    □

Next we consider eigenfunctions of trees which vanish on some vertices.

**Theorem 4.5 ([19]).** *Let* $\mathbf{M}$ *be a generalized Laplacian of a tree* $T$ *on* $n$ *vertices and let* $\lambda$ *be an eigenvalue of* $\mathbf{M}$ *with eigenfunction* $f$ *that vanishes on some vertices. Then there exists an ordering of the vertices such that* $\mathbf{M}$ *has the form*

$$\mathbf{M} = \begin{pmatrix} \mathbf{M}_1 & 0 & \cdots & 0 & \mathbf{B}_1^\mathsf{T} \\ 0 & \mathbf{M}_2 & \cdots & 0 & \mathbf{B}_2^\mathsf{T} \\ \vdots & \vdots & \ddots & \vdots & \vdots \\ 0 & 0 & \cdots & \mathbf{M}_m & \mathbf{B}_m^\mathsf{T} \\ \mathbf{B}_1 & \mathbf{B}_2 & \cdots & \mathbf{B}_m & \mathbf{M}_Z \end{pmatrix} \tag{4.1}$$

*and the following statements hold:*

  (i) $\lambda$ *is a simple eigenvalue of all block matrices* $\mathbf{M}_1, \ldots, \mathbf{M}_m$ *and each matrix* $\mathbf{M}_j$ *has an eigenfunction of* $\lambda$ *that does not vanish on any vertex.*
 (ii) *Let* $k_1, \ldots, k_m$ *be the positions of* $\lambda$ *in the spectra of* $\mathbf{M}_1, \ldots, \mathbf{M}_m$, *resp., in nondecreasing order. Then the number of strong nodal domains of an eigenfunction* $f$ *of* $\mathbf{M}$ *with eigenvalue* $\lambda$ *is at most* $k_1 + \cdots + k_m$.
(iii) *There exists an eigenfunction* $f$ *of* $\mathbf{M}$ *with eigenvalue* $\lambda$ *with* $k_1 + \cdots + k_m$ *strong nodal domains. Such an eigenfunction can be found in* $\mathcal{O}(n^2)$ *time.*

*Proof.* Let $Z$ be the set of all vertices where every eigenfunction with eigenvalue $\lambda$ vanishes. If $\lambda$ is simple then $Z$ is obviously well defined and nonempty by the assumption of the theorem.

Now suppose that the multiplicity of the eigenvalue $\lambda$ is $r \geq 2$. Then each eigenfunction $f$ with eigenvalue $\lambda$ can be expressed as a linear combination of a basis $g_1, \ldots, g_r$ of the eigenspace $\mathcal{E}_\lambda$. By Lemma 4.3 each of these linear combinations must vanish on some vertex $v$. The set $H_v$ of linear combinations of $g_1, \ldots, g_r$ that vanish on a particular vertex $v$ is either a hyperplane (i.e., a subspace with codimension 1) in the eigenspace $\mathcal{E}_\lambda$ or equals $\mathcal{E}_\lambda$ itself (see also Sect. 5.1). The union of all these subspaces $H_v$ is then the set of all linear combinations that vanish on at least one vertex. By Lemma 4.3 this union equals the eigenspace $\mathcal{E}_\lambda$ and thus $\mathcal{E}_\lambda = H_w$ for at least one vertex $w$. Thus $w \in Z$, i.e., $Z$ is not empty. We observe that $Z$ is unique defined and does not depend on the particular choice of $g_1, \ldots, g_r$.

The graph $G - Z$ is a forest with components $T_1, \ldots, T_m$. Let $\mathbf{M}_1, \ldots, \mathbf{M}_m$ be the generalized Laplacians restricted to $T_1, \ldots, T_m$. Since there are no edges between two vertices $v_i$ and $v_j$ that lie in different subtrees $T_i$ and $T_j$, $i \neq j$, there exists an ordering of the vertices such that $\mathbf{M}$ has the form (4.1).

($i$) Let $f$ be some eigenfunction of $\mathbf{M}$ corresponding to eigenvalue $\lambda$. For each vertex $v \in T_j$ we have $\lambda f(v) = (\mathbf{M}f)(v) = \sum_{u \in V(G)} M_{vu} f(u) = \sum_{u \in V(T_j)} M_{vu} f(u) = \sum_{u \in V(T_j)} (M_j)_{vu} f(u)$ because $M_{vu} = 0$ for all $u \notin V(T_j) \cup Z$ and $f(u) = 0$ for all $u \in Z$. Therefore $\lambda$ is an eigenvalue of the matrices $\mathbf{M}_j$ with $f$ restricted to $T_j$ as eigenfunction.

Above we have chosen $f$ from $\mathcal{E}_\lambda$ without further restrictions. We can in particular select an eigenfunction $f$ that is nonzero on every vertex in $T_j$. Otherwise, by Lemma 4.3 and the same argument as above there would be a vertex $w$ in $T_j$ where all $f$ vanish. Thus $w \in Z$ contradicting the assumption that $T_j$ is a component of $G - Z$. Hence there is an eigenfunction of eigenvalue $\lambda$ which is nonzero on every vertex of $G - Z$ and thus by Thm. 4.4, $\lambda$ is a simple eigenvalue of $\mathbf{M}_j$.

($ii$) From these considerations it follows that all eigenfunctions $f$ restricted to a component $T_j$ are linearly dependent (as $\lambda$ is a simple eigenvalue on $T_j$) and must be nonzero on every vertex in $T_j$ (as otherwise a zero vertex belongs to $Z$). By Thm. 4.4, the restriction of an eigenfunction $f$ of $\mathbf{M}(G)$ to $T_j$ has exactly $k_j$ strong nodal domains. Therefore, $\mathfrak{S}(f) \leq k_1 + \cdots + k_m$.

($iii$) By the above the set of all eigenfunctions of eigenvalue $\lambda$ restricted to a component $T_j$ is a vector space of dimension 1 that is spanned by some nonzero function $b_j$. Then every eigenfunction $f$ with eigenvalue $\lambda$ can be written as $f(v) = \sum_{j=1}^m \beta_j \, b_j(v)$. There must exist an eigenfunction where no $\beta_j$ is zero, since by the same arguments as above $Z$ would contain vertices from some component $T_j$, a contradiction. If $\lambda$ is not a simple eigenvalue then such a combination can be found by the following algorithm:

    compute basis $g_1, \ldots, g_r$ for eigenspace $\mathcal{E}_\lambda$;
    $f := g_1$;
    **for** $i = 2, \ldots, r$ **do**
        $f := f + \alpha_i \, g_i$, choose $\alpha_i$ s.t.
        $\alpha_i \neq 0$ and $\alpha_i \notin \left\{ -\frac{f(x)}{g_i(x)} : g_i(x) \neq 0, \, x \in V(G) \right\}$.

For the resulting function $f$ we have $\mathfrak{S}(f) = \mathfrak{S}(b_1) + \cdots + \mathfrak{S}(b_m) = k_1 + \cdots + k_m$, as claimed. It is easy to see that we need $\mathcal{O}(n^2)$ operations to find this eigenfunction $f$ from an arbitrary eigensystem of $\mathbf{M}$.    □

We remark that it is an open problem to find the maximum (or minimum) number of weak nodal domains of an eigenfunction for a given eigenvalue of a generalized Laplacian of a tree.

### 4.1.2 The Minimum Number of Nodal Domains

In Sect. 4.1.1 we have seen, that it is not to difficult to compute an eigenfunction of an eigenvalue $\lambda$ of a given generalized Laplacian $\mathbf{M}$ of a tree that has the maximum possible number of strong nodal domains. Finding the minimum number of strong nodal domain is much more difficult; in fact we have to solve the following problem which turns out to be NP-complete.

**Problem 4.6** (MINIMUM NUMBER OF NODAL DOMAINS).
*Instance:* A generalized Laplacian matrix $\mathbf{M}$ of a tree, an eigenvalue $\lambda$ of $\mathbf{M}$ with multiplicity $r \geq 2$.
*Question:* Find an eigenfunction $f$ of $\lambda$ such that the number $\mathfrak{S}(f)$ of strong nodal domains of is minimal.

Let $\mathbf{M}$ be a generalized Laplacian of a tree with an eigenvalue $\lambda$ of multiplicity $r \geq 2$. In the proof of Thm. 4.5 we have seen that there exists a nonempty set $Z$ of vertices where all eigenfunctions $f$ of $\lambda$ vanish. Furthermore, every such eigenfunction can be decomposed as $f = \sum_{j=1}^{m} \beta_j\, b_j$ where the functions $b_j$ have nonzero vertices in the component $T_j$ of $G - Z$ and vanish outside $T_j$. In particular this holds for each function $g_1, \ldots, g_r$ of a basis of the eigenspace $\mathcal{E}_\lambda$ and we have $g_i = \sum_{j=1}^{m} \beta_{ji}\, b_j$, for $i = 1, \ldots, r$. Now let $\mathbf{B} = (\beta_{ji})$, $j = 1, \ldots, m$, $i = 1, \ldots, r$. Then every eigenfunction $f$ can be computed as $f = \sum_{i=1}^{r} a_i\, g_i = \sum_{j=1}^{m} \left( \sum_{i=1}^{r} a_i \beta_{ji} \right) b_i = \sum_{j=1}^{m} (\mathbf{Ba})_i\, b_i$. for some coefficients $\mathbf{a} = (a_1, \ldots, a_r)$. We introduce a new function $c_i(\mathbf{a})$ by $c_i(\mathbf{a}) = 1$ if $(\mathbf{Ba})_i \neq 0$, and $c_i(\mathbf{a}) = 0$ otherwise. Hence the number of nodal domains of $f$ is given by $\mathfrak{S}(f) = k_1\, c_1(\mathbf{a}) + \cdots + k_m\, c_m(\mathbf{a})$, where $k_1, \ldots, k_m$ is defined as in Thm. 4.5. Consequently, we have to solve the minimization problem $k_1 c_1(\mathbf{a}) + \cdots + k_m c_m(\mathbf{a})$ for nonzero vectors $\mathbf{a}$. Therefore, the decision problem of MINIMUM NUMBER OF NODAL DOMAINS is the following problem.

**Problem 4.7** (MIN($\mathfrak{S}$)).
*Instance:* An $(m \times r)$ matrix $\mathbf{B}$ with real entries, positive integers $k_1, \ldots, k_m$, and a positive integer $s$.
*Question:* Is there a nonzero rational vector $\mathbf{x} = (x_1, \ldots, x_r)$, such that $k_1 c_1(\mathbf{x}) + \cdots + k_m c_m(\mathbf{x}) \leq s$?

The matrix $\mathbf{B}$ of this decision problem can be arbitrary large. As an examples take a binary tree with $n = 2^{q+1} - 1$ vertices and $k = 2^q$ leaves (nodes of degree 1). Then by a result of Faria [62] this tree has eigenvalue $\lambda = 1$ with multiplicity $r \geq 2^q - 2^{q-1} = 2^{q-1} \geq \frac{n}{4}$. Obviously $m$ is at least the number of leaves and thus $m \geq \frac{n}{2}$.

We have to reduce a known NP-complete problem to $\mathsf{MIN}(\mathfrak{S})$ to show that the latter problem is also NP-complete. For this task we use the MINIMUM SUPPORT problem for the reduction.

**Problem 4.8 (MINIMUM SUPPORT).**
*Instance:* An $(m \times r)$ matrix $\mathbf{B}$ with rational entries, a positive integer $s$.
*Question:* Is there a nonzero rational vector $\mathbf{x} = (x_1, \ldots, x_r)$ such that the number of nonzero elements of $\mathbf{Bx}$ is at most $s$, i.e., support$(\mathbf{Bx}) \leq s$?

MINIMUM SUPPORT is NP-complete, see e.g. [19]. By a straightforward computation we can reduce MINIMUM SUPPORT to $\mathsf{MIN}(\mathfrak{S})$. Thus we have the following result.

**Theorem 4.9 ([19]).** *The decision problem* $\mathsf{MIN}(\mathfrak{S})$ *is NP-complete.*

## 4.2 Cographs and Threshold Graphs

In the previous section we have seen that it is not always easy to find the maximum or minimum number of nodal domains for a particular graph class. In this section we see that this problem is easy for cographs.

$G$ is a complement-reducible graph (*cograph*) if the complement of every nontrivial connected subgraph of $G$ is disconnected. Cographs arise in many disparate areas of mathematics and computer science and have several characterizations, see e.g. [42] for a short survey. For instance, $G$ is a cograph if and only if it has no induced subgraph $P_4$ (a path with 4 vertices). For our purpose the following characterization in terms of join and disjoint union operations will be particularly useful:

Let $G_1(V_1, E_1)$ and $G_2(V_2, E_2)$ be graphs on disjoint sets of $r$ and $s$ vertices, respectively. Their *disjoint union* $G_1 + G_2$ is the graph $G_1 + G_2 = (V_1 \cup V_2, E_1 \cup E_2)$, and their *join* $G_1 * G_2$ is the graph on $n = r + s$ vertices obtained from $G_1 + G_2$ by inserting new edges from each vertex of $G_1$ to each vertex of $G_2$.

**Proposition 4.10 ([43]).** *To each cograph $G(V, E)$ one can associate a unique rooted tree $T$, called the* cotree *of $G$. Each leaf node of $T$ corresponds to a (unique) vertex of $V$. Each internal node is labeled with either a '$*$' or a '$+$'. Children of nodes labeled with '$+$' are labeled with '$*$', and vice versa. It is possible to associate a cograph with each node of the cotree $T$. Leaf nodes correspond to the cograph with the one vertex they represent. Internal nodes labeled with '$*$' ('$+$') correspond to the join (disjoint union) of the cographs, corresponding to the children of the node (see Fig. 4.1). $G$ equals the cograph corresponding with the root of $T$. Cographs can be recognized in $\mathcal{O}(|V| + |E|)$ time, and in the same time the corresponding cotree can be built.*

This proposition states that each cograph $G$ either is the disjoint union of two disjoint cographs $G_1$ and $G_2$, $G = G_1 + G_2$, or $G$ is the join of two disjoint cographs $G_1$ and $G_2$, $G = G_1 * G_2$.

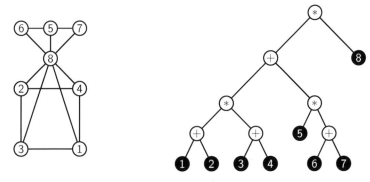

**Fig. 4.1.** Cograph (l.h.s.) and its cotree (r.h.s.)

The following lemma can be used to investigate the nodal domains of cographs.

**Proposition 4.11 ([134]).** *Let $G_1$ and $G_2$ be graphs on disjoint sets of $r$ and $s$ vertices, respectively. If $\mu_1 \leq \cdots \leq \mu_r$ and $\nu_1 \leq \cdots \leq \nu_s$ are eigenvalues of the graph Laplacian of $G_1$ and $G_2$, respectively. Then the eigenvalues of $G_1 * G_2$ are $n = r + s$; $\mu_2 + s, \ldots, \mu_r + s$; $\nu_2 + r, \ldots, \nu_s + r$; and 0.*

*Suppose $f$ is an eigenfunction of $G_1$ that is orthogonal to $\mathbf{1}$. Extend $f$ to $G_1 * G_2$ by defining it to be zero on $V(G_2)$. If $f$ affords the eigenvalue $\mu$, the extension of $f$ is an eigenfunction of $G_1 * G_2$ affording $\mu + s$. Similarly an eigenfunction of $G_2$ affording $\nu$ extends to an eigenfunction of $G_1 * G_2$ affording $\nu + r$. The eigenvalue $\lambda = r + s$ corresponds to an eigenfunction whose value is $-s$ on each of the $r$ vertices of $G_1$ and $r$ on each of the $s$ vertices of $G_2$. Finally, the trivial eigenvalue is afforded by the constant function $\mathbf{1}$ on $V_1 \cup V_2$.*

Obviously, the eigenvalues of the Laplacian $\mathbf{L}(G_1 + G_2)$ are the union of eigenvalues of $\mathbf{L}(G_1)$ and $\mathbf{L}(G_2)$ (respecting multiplicity). It follows from Props. 4.10 and 4.11 that the Laplacian eigenvalues of a cograph are integers. Moreover, we can directly compute the number of weak nodal domains of a cograph.

**Corollary 4.12.** *Let $G$ be a connected cograph with $n$ vertices. $G$ has the form $G = G_1 * G_2$. Let $c_1 \leq c_2$ be the number of components of $G_1$ and $G_2$, respectively.*

*(i) Every eigenfunction $f$ affording the eigenvalue $0 < \lambda < n$ of the Laplacian $\mathbf{L}(G)$ has $\mathfrak{W}(f) = 2$ weak nodal domains.*

*(ii) If $c_1 \geq 2$, then the largest eigenvalue $\lambda = n$ is simple and its eigenfunction has two weak (strong) nodal domains. If $c_1 = 1$, then every eigenfunction $f$ of $\lambda = n$ has either two weak (strong) nodal domains or $\mathfrak{S}(f) = \mathfrak{W}(f) = c_2 + 1$.*

**Theorem 4.13.** *For each eigenvalue of the Laplacian of a cograph $G(V, E)$ we can find an eigenfunction with maximum or minimum number of strong nodal domains in $\mathcal{O}(|V| + |E|)$ time.*

*Proof.* For a rooted tree $T$ and a vertex $v \in V(T)$ the *subtree at* $v$ is the induced tree by $v$ and all descendants of $v$. Similarly, a *subtree of* $v$ is the subtree at one of the children of $v$. In the following we denote by $\mathfrak{S}_{\max}(G_i, \lambda)$ and $\mathfrak{S}_{\min}(G_i, \lambda)$ the maximum and minimum number of strong nodal domains of an eigenfunction corresponding to eigenvalue $\lambda$ in a graph $G_i$, respectively.

Now by Prop. 4.10 a cograph $G$ has a unique cotree $T$. Let $v$ be some node of the cotree $T$ with subtrees $T_1, \ldots, T_k$, and let $G_1, \ldots, G_k$ be the corresponding cographs to these cotrees. Let $G_v$ be the cograph corresponding with $v$ as root. We show that the number of strong nodal domains of $G_v$ can be expressed in terms of the number of strong nodal domains of $G_1, \ldots, G_k$. For a function $f_i$ on a subgraph $G_i$ we use the symbol $\hat{f}_i$ for the function extended to $G_v$ with $\hat{f}_i(x) = f_i(x)$ for all $x \in G_i$ and $\hat{f}_i(x) = 0$ otherwise.

Suppose $v$ has label '+' (disjoint union), then the eigenvalues of $G_v$ are the union of eigenvalues of $G_1, \ldots, G_k$. Let $f_1, \ldots, f_k$ be eigenfunctions of an eigenvalue $\lambda$ with maximum number of strong nodal domains, i.e. $\mathfrak{S}(f_i) = \mathfrak{S}_{\max}(G_i, \lambda)$. Then $f = \sum_{i=1}^{k} \hat{f}_i$ is an eigenfunction of $\lambda$ of the cograph $G_v$ with maximum number of strong nodal domains and $\mathfrak{S}_{\max}(G_v, \lambda) = \mathfrak{S}(f) = \sum_{i=1}^{k} \mathfrak{S}(f_i) = \sum_{i=1}^{k} \mathfrak{S}_{\max}(G_i, \lambda)$. Similarly let $g_1, \ldots, g_k$ be eigenfunctions of $\lambda$ with minimum number of nodal domains. Then each $\hat{g}_i$ is an eigenfunction of $\lambda$ in $G_v$ and thus $\mathfrak{S}_{\min}(G_v, \lambda) = \min_{i=1, \ldots, k} \mathfrak{S}_{\min}(G_i, \lambda)$.

Now assume $v$ has label '$*$' (join operation), then an easy induction gives that the eigenvalues of $G_v$ are $|V(G_v)|$ and $\lambda_{G_i} + \sum_{j \neq i} |V(G_j)|$, for $i = 1, \ldots, k$, where $\lambda_{G_i} > 0$ is an eigenvalue of $G_i$. By Prop. 4.11 the extension $\hat{f}_i$ of an eigenfunction $f_i$ of $\lambda_{G_i}$ is an eigenfunction of $\mu = \lambda_{G_i} + \sum_{j \neq i} |V(G_j)|$. The extended eigenfunctions $\hat{f}_{i_1}, \ldots, \hat{f}_{i_p}$ that afford the same eigenvalue $\mu \neq |V(G_v)|$ span the eigenspace of $\mu$; all of them have at least two vertices with opposite sign, since $\mu$ cannot be the first (smallest) eigenvalue. By the join operation all linear combinations of at least two of these functions have two strong nodal domains. Therefore $\mathfrak{S}_{\max}(G_v, \mu) = \max_{i=1, \ldots, k} \mathfrak{S}_{\max}(G_i, \mu)$ and $\mathfrak{S}_{\min}(G_v, \mu) = 2$ if the eigenspace to eigenvalue $\mu$ has multiplicity $p \geq 2$; and $\mathfrak{S}_{\min}(G_v, \mu) = \mathfrak{S}_{\min}(G_{i_1}, \mu)$ if $p = 1$.

For the eigenvalue $\mu = |V(G_v)|$ the children of the node $v$ are labeled with '+' by Prop. 4.10. Therefore each of the graphs $G_1, \ldots, G_k$ is either disconnected or a single vertex. Let $c_1, \ldots, c_k$ be the number of connected components of $G_1, \ldots, G_k$. By Prop. 4.11 it is easy to see that the maximum number of nodal domains for eigenvalue $\mu$ is given by $\max\{c_1, \ldots, c_k\} + 1$ when the node $v$ has more than two children and given by $c_1 + c_2$ when $v$ has two children.

We have shown that it is enough to build the cotree of a cograph to find the maximum and minimum number of strong nodal domains. Proposition 4.10

guarantees that we can build the cotree of a cograph $G(V, E)$ in $\mathcal{O}(|V| + |E|)$ time. □

**Corollary 4.14.** *The Laplacian eigenvalues of a complete $k$-partite graph $K_{n_1,\ldots,n_k}$ with $n_1 \geq \cdots \geq n_k$ are $0$; $n = n_1 + \cdots + n_k$; and $n - n_i$, for $i = 1, \ldots, k$. The maximum number of strong nodal domains of eigenvalues $n$ and $n - n_i$ are equal to $n_1 + n_2$ and $n_i$, respectively. The minimum number of nodal domains of all eigenvalues equals two.*

For an important subclass of cographs, namely threshold graphs, we can directly compute the number of nodal domains without using Thm. 4.13. A graph $G(V, E)$ is called as a *threshold graph*, if $G$ does not contain one of the three forbidden induced subgraphs: $K_2 + K_2$, $C_4$, or $P_4$. Another useful characterization of threshold graph is given by the following result.

**Proposition 4.15 ([36]).** *A graph $G$ is a connected threshold graph if and only if $G = (K, U)$, where $K$ is a complete graph with a partition of nonempty cliques $K_1, \ldots, K_s$ and $U$ is an independent set of vertices with a partition of nonempty independent sets $U_1, \ldots, U_s$. All vertices of $K_i$ are adjacent to all vertices of $U_h$, for $i = 1, \ldots, s$ and $1 \leq h \leq i$, see Fig. 4.2.*

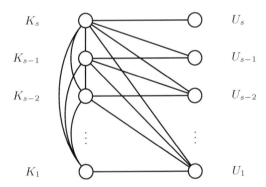

**Fig. 4.2.** The typical structure of a threshold graph. A line between cells $K_i$ and $U_j$ indicates that each vertex in $K_i$ is adjacent to each vertex of $U_j$.

The Laplacian eigenvalues of a threshold graph are obtained easily by induction using Props. 4.11 and 4.15, see also [93].

**Theorem 4.16.** *Let $G = (K, U)$ be a connected threshold graph with the partitions $K_i$ and $U_i$, for $i = 1, \ldots, s$. The eigenvalues of the Laplacian $\mathbf{L}(G)$ are $0$; $\sum_{i=1}^{h} |U_i| + \sum_{j=1}^{s} |K_j|$ for $h = 1, \ldots, s$; $\sum_{j=h}^{s} |K_j|$ for $h = 2, \ldots, s$; and $\sum_{j=1}^{s} |K_j|$ when $|U_1| \geq 2$. The bounds for the number of strong nodal domains are given as follows.*

(i) If $\lambda = \sum_{i=1}^{h} |U_i| + \sum_{j=1}^{s} |K_j|$, then
  (a) $2 \leq \mathfrak{S}(f) \leq |U_h| + 2$ when $h \geq 2$,
  (b) $2 \leq \mathfrak{S}(f) \leq |U_1| + 1$ when $h = 1$.

(ii) If $\lambda = \sum_{j=h}^{s} |K_j|$, then
  (a) $2 \leq \mathfrak{S}(f) \leq |U_h| + 1$ when $h \geq 2$,
  (b) $\mathfrak{S}(f) \leq |U_1|$ when $h = 1$ and $|U_1| \geq 2$.

These bounds on $\mathfrak{S}(f)$ are sharp.

## 4.3 Product Graphs and the Boolean Hypercube

In this section we study nodal domains of Boolean hypercubes. These are special cases of *graph products*. Given two nonempty graphs $G(V, E)$ and $H(W, F)$. Then the *Cartesian product* $G \square H$ is a graph with vertex set $V \times W$ and $(x_1, x_2)(y_1, y_2)$ is an edge in $E(G \square H)$ if and only if either $x_2 = y_2$ and $x_1 y_1 \in E(G)$ or if $x_1 = y_1$ and $x_2 y_2 \in E(H)$. One may view $G \square H$ as the graph obtained from $G$ by replacing each of its vertices with a copy of $H$ and each of its edges with $|V(H)|$ edges joining corresponding vertices of $H$ in the two copies. The graph product is a commutative, associative binary operation on graphs, see e.g. [102].

The *Kronecker product*, also known as *tensor product* or *direct product*, of two matrices $\mathbf{A}$ and $\mathbf{B}$ of respective sizes $m \times n$ and $s \times t$ is the $ms \times nt$ partitioned matrix

$$\mathbf{A} \otimes \mathbf{B} = \begin{pmatrix} a_{11}B & a_{12}B & \cdots & a_{1n}B \\ a_{21}B & a_{22}B & \cdots & a_{2n}B \\ \vdots & \vdots & \vdots & \vdots \\ a_{m1}B & a_{m2}B & \cdots & a_{mn}B \end{pmatrix} .$$

Let $G$ and $H$ be graphs with $n$ and $s$ vertices, respectively. The Laplacian matrix of the Cartesian product is then given by

$$\mathbf{L}(G \square H) = \mathbf{L}(G) \otimes \mathbf{I}_s + \mathbf{I}_n \otimes \mathbf{L}(H) .$$

If $f$ is an eigenfunction of $\mathbf{L}(G)$ affording the eigenvalue $\lambda$ and $g$ an eigenfunction of $\mathbf{L}(H)$ affording the eigenvalue $\mu$, then $f \otimes g$ is an eigenfunction of $\mathbf{L}(G \square H)$ affording the eigenvalue $\lambda + \mu$, where $(f \otimes g)(x, y) = f(x) g(y)$ for $(x, y) \in V \times W$. Therefore the eigenvalues of $\mathbf{L}(G \square H)$ is the set of sums of the eigenvalues of $\mathbf{L}(G)$ and $\mathbf{L}(H)$, see e.g. [134]. By the definitions of $G \square H$ and the Kronecker product,

$$\mathfrak{S}(f \otimes g) \leq \mathfrak{S}(f) \mathfrak{S}(g) \qquad \text{and} \qquad \mathfrak{W}(f \otimes g) \leq \mathfrak{W}(f) \mathfrak{W}(g) .$$

Notice, however, that it is not possible to derive an upper or lower bound on the number of nodal domains for eigenfunctions on the $\mathbf{L}(G \square H)$ from the corresponding bounds of $G$ and $H$, as the following example shows. The graph $G$ in Fig. 4.3 has the eigenvalues $\lambda_3(G) = 1$ (of multiplicity 2) and $\lambda_5(G) = 3$ (simple). The respective eigenfunctions do not have more than 3 and 2 strong nodal domains. However, there exists an eigenfunction of $G \square K_2$ to eigenvalue $\lambda_7(G \square K_2) = 3 + 0 = 1 + 2$ with 8 strong nodal domains (Fig. 4.3, r.h.s.), whereas we only get $3 \times 2 = 6$ and $2 \times 1 = 2$ nodal domains for eigenfunctions that are products of corresponding eigenfunctions of $G$ and $K_2$, respectively.

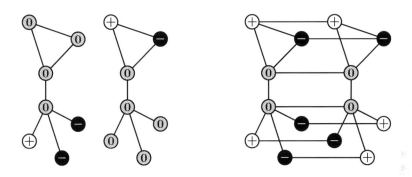

**Fig. 4.3.** Graph $G$ (l.h.s.) and product graph $G \square K_2$ (r.h.s.). The sign patterns of eigenfunctions with maximal number of nodal domains are shown for eigenvalues $\lambda_3(G) = 1$, $\lambda_5(G) = 3$ and $\lambda_7(G \square K_2) = 3$, respectively. Notice, that $\mathfrak{S}(f_7) \not\leq \mathfrak{S}(f_3)\,\mathfrak{S}(f_5)$.

Similar counterexamples for the number of weak nodal domains exist but are more complicated. Nevertheless, we have found such examples by numerical experiments, see results for the hypercube in Table 5.1 on p. 73.

### 4.3.1 The Boolean Hypercube

The *hypercube* $K_2^N$ is the graph with vertex set $V = \{(v_1, v_2, \ldots, v_N)|v_i = \pm 1\}$ and edges connecting two vertices that differ in a single coordinate, i.e., $uv \in E$ if and only if $u_i = v_i$ for all but one index $j$ for which we then have $u_j = -v_j$. The number $N$ of coordinates is usually called the *dimension* of $K_2^N$. The graph $K_2^N$ has $|V| = 2^N$ vertices and $|E| = N\,2^{N-1}$ edges. It is not hard to verify that the hypercube is a bipartite graph and it is equivalently defined as $N$-fold Cartesian product of $K_2$, the graph consisting of a single edge and its two end vertices. The *Walsh functions* [69, 171]

$$\varphi_I(v) = \prod_{k \in I} v_k \tag{4.2}$$

where $I \subseteq \{1, 2, \ldots, N\}$ form a complete set of eigenfunctions of the Laplacian of the hypercube. A short direct computation verifies that these functions satisfy the eigenvalue equation

$$\mathbf{L}\varphi_I = 2|I|\,\varphi_I$$

and the orthogonality relation

$$\langle \varphi_I, \varphi_J \rangle = \sum_{v \in V} \varphi_I(v)\varphi_J(v) = \delta_{I,J}|V|$$

where the $\delta_{I,J} = 1$ if $I = J$ and 0 otherwise. Thus there are $\binom{N}{|I|}$ eigenfunctions with eigenvalue $2|I|$. It is customary to call $p = |I|$ the order of the Walsh function $\varphi_I$. As a consequence we can write every eigenfunction $f$ corresponding to eigenvalue $\lambda = 2p$ as

$$f(v) = \sum_{I:\,|I|=p} a_I\,\varphi_I(v) \tag{4.3}$$

where $a_I \in \mathbb{R}$. The Walsh functions satisfy the following important recursion w.r.t. the number $N$ of coordinates:

$$\varphi_I'(v; v_{N+1}) = \varphi_I(v) \qquad \text{and} \qquad \varphi_{I \cup \{N+1\}}'(v; v_{N+1}) = v_{N+1}\,\varphi_I(v) . \tag{4.4}$$

It is sometime more convenient to write (4.4) as a tensor product:

$$\varphi_I' = \begin{pmatrix} 1 \\ 1 \end{pmatrix} \otimes \varphi_I \qquad \text{and} \qquad \varphi_{I \cup \{N+1\}}' = \begin{pmatrix} 1 \\ -1 \end{pmatrix} \otimes \varphi_I .$$

Clearly, $\varphi_{I \cup \{N+1\}}'$ is an eigenfunction of $K_2^{N+1}$ with eigenvalues $2(|I|+1)$. It follows that all Walsh functions can be obtained recursively in this way. Alternatively we may view the Boolean hypercube as Cayley graph over the Abelian group $(\{0,1\}, \oplus)^N$; as discussed in section 2.7, the Walsh functions therefore are essentially the irreducible representations of this group. More details and further applications of the group theoretical perspective are discussed in [45, 58].

Equation (4.4) of course holds for any eigenfunction $f$ of $K_2^N$ with eigenvalue $2p$:

$$f^+ = \begin{pmatrix} 1 \\ 1 \end{pmatrix} \otimes f \qquad \text{and} \qquad f^- = \begin{pmatrix} 1 \\ -1 \end{pmatrix} \otimes f \tag{4.5}$$

are eigenfunctions of $K_2^{N+1}$ with eigenvalues $2p$ and $2(p+1)$, respectively.

### 4.3.2 The Number of Nodal Domains

It follows immediately from Thm. 3.1 that an eigenfunction $f$ with eigenvalue $2p$ has at most

$$\mathfrak{S}(f) \le \mathfrak{s}_{N,p} = \sum_{j=0}^{p} \binom{N}{j} \qquad \text{and} \qquad \mathfrak{W}(f) \le \mathfrak{w}_{N,p} = 1 + \sum_{j=0}^{p-1} \binom{N}{j} \tag{4.6}$$

**Table 4.1.** Upper bounds on the number of strong and weak nodal domains as function of $N$ and $p = |I|$ as given in (4.6) and (4.7), respectively.

$\mathfrak{s}_{N,p}$

| $p=$ | 0 | 1 | 2 | 3 | 4 | 5 | 6 | 7 | 8 | 9 | 10 | 11 | 12 | 13 | 14 |
|---|---|---|---|---|---|---|---|---|---|---|---|---|---|---|---|
| $N$ | | | | | | | | | | | | | | | |
| 2 | 1 | 3 | 4 | | | | | | | | | | | | |
| 3 | 1 | 4 | 7 | 8 | | | | | | | | | | | |
| 4 | 1 | 5 | 11 | 15 | 16 | | | | | | | | | | |
| 5 | 1 | 6 | 16 | 26 | 31 | 32 | | | | | | | | | |
| 6 | 1 | 7 | 22 | 42 | 57 | 63 | 64 | | | | | | | | |
| 7 | 1 | 8 | 29 | 64 | 99 | 120 | 127 | 128 | | | | | | | |
| 8 | 1 | 9 | 37 | 93 | 163 | 219 | 247 | 255 | 256 | | | | | | |
| 9 | 1 | 10 | 46 | 130 | 256 | 382 | 466 | 502 | 511 | 512 | | | | | |
| 10 | 1 | 11 | 56 | 176 | 386 | 638 | 848 | 968 | 1013 | 1023 | 1024 | | | | |
| 11 | 1 | 12 | 67 | 232 | 562 | 1024 | 1486 | 1816 | 1981 | 2036 | 2047 | 2048 | | | |
| 12 | 1 | 13 | 79 | 299 | 794 | 1586 | 2510 | 3302 | 3797 | 4017 | 4083 | 4095 | 4096 | | |
| 13 | 1 | 14 | 92 | 378 | 1093 | 2380 | 4096 | 5812 | 7099 | 7814 | 8100 | 8178 | 8191 | 8192 | |
| 14 | 1 | 15 | 106 | 470 | 1471 | 3473 | 6476 | 9908 | 12911 | 14913 | 15914 | 16278 | 16369 | 16383 | 16384 |

$\mathfrak{w}_{N,p}$

| $p=$ | 0 | 1 | 2 | 3 | 4 | 5 | 6 | 7 | 8 | 9 | 10 | 11 | 12 | 13 | 14 |
|---|---|---|---|---|---|---|---|---|---|---|---|---|---|---|---|
| $N$ | | | | | | | | | | | | | | | |
| 2 | 1 | 2 | 4 | | | | | | | | | | | | |
| 3 | 1 | 2 | 5 | 8 | | | | | | | | | | | |
| 4 | 1 | 2 | 6 | 12 | 16 | | | | | | | | | | |
| 5 | 1 | 2 | 7 | 17 | 27 | 32 | | | | | | | | | |
| 6 | 1 | 2 | 8 | 23 | 43 | 58 | 64 | | | | | | | | |
| 7 | 1 | 2 | 9 | 30 | 65 | 100 | 121 | 128 | | | | | | | |
| 8 | 1 | 2 | 10 | 38 | 94 | 164 | 220 | 248 | 256 | | | | | | |
| 9 | 1 | 2 | 11 | 47 | 131 | 257 | 383 | 467 | 503 | 512 | | | | | |
| 10 | 1 | 2 | 12 | 57 | 177 | 387 | 639 | 849 | 969 | 1014 | 1024 | | | | |
| 11 | 1 | 2 | 13 | 68 | 233 | 563 | 1025 | 1487 | 1817 | 1982 | 2037 | 2048 | | | |
| 12 | 1 | 2 | 14 | 80 | 300 | 795 | 1587 | 2511 | 3303 | 3798 | 4018 | 4084 | 4096 | | |
| 13 | 1 | 2 | 15 | 93 | 379 | 1094 | 2381 | 4097 | 5813 | 7100 | 7815 | 8101 | 8179 | 8192 | |
| 14 | 1 | 2 | 16 | 107 | 471 | 1472 | 3474 | 6477 | 9909 | 12912 | 14914 | 15915 | 16279 | 16370 | 16384 |

$\mathfrak{w}^*_{N,p}$

| $p=$ | 0 | 1 | 2 | 3 | 4 | 5 | 6 | 7 | 8 | 9 | 10 | 11 | 12 | 13 | 14 |
|---|---|---|---|---|---|---|---|---|---|---|---|---|---|---|---|
| $N$ | | | | | | | | | | | | | | | |
| 2 | 1 | 2 | 4 | | | | | | | | | | | | |
| 3 | 1 | 2 | 4 | 8 | | | | | | | | | | | |
| 4 | 1 | 2 | 4 | 10 | 16 | | | | | | | | | | |
| 5 | 1 | 2 | 4 | 12 | 24 | 32 | | | | | | | | | |
| 6 | 1 | 2 | 4 | 14 | 34 | 54 | 64 | | | | | | | | |
| 7 | 1 | 2 | 4 | 16 | 46 | 86 | 116 | 128 | | | | | | | |
| 8 | 1 | 2 | 4 | 18 | 60 | 130 | 200 | 242 | 256 | | | | | | |
| 9 | 1 | 2 | 4 | 20 | 76 | 188 | 328 | 440 | 496 | 512 | | | | | |
| 10 | 1 | 2 | 4 | 22 | 94 | 262 | 514 | 766 | 934 | 1006 | 1024 | | | | |
| 11 | 1 | 2 | 4 | 24 | 114 | 384 | 774 | 1278 | 1698 | 1938 | 2028 | 2048 | | | |
| 12 | 1 | 2 | 4 | 26 | 136 | 466 | 1126 | 2050 | 2974 | 3634 | 3964 | 4074 | 4096 | | |
| 13 | 1 | 2 | 4 | 28 | 160 | 600 | 1590 | 3174 | 5022 | 6606 | 7596 | 8036 | 8168 | 8192 | |
| 14 | 1 | 2 | 4 | 30 | 186 | 758 | 2188 | 4762 | 8194 | 11626 | 14200 | 15630 | 16202 | 16358 | 16384 |

strong and weak nodal domains, respectively. Numerical values are listed in Table 4.1.

These upper bounds are not sharp and despite the "simplicity" of both the graph class and the basis of the eigenspace it is not trivial to compute the possible number of nodal domains (see Chap. 5). Indeed, we only present a few partial results as many problems remain open.

For the Walsh functions (4.2) one easily can show that $\mathfrak{S}(\varphi_I) = \mathfrak{W}(\varphi_I) = 2^{|I|}$. Thus from representation (4.3) we immediately have the following result.

**Theorem 4.17.** *The eigenvalue $2p$ has an associated eigenfunction $f$ with at least $2^p$ nodal domains, $\mathfrak{S}(f) \geq \mathfrak{W}(f) \geq 2^p$, for all $0 \leq p \leq N$.*

On the other hand we can use some symmetry relations to improve the nodal domain theorem for Boolean hypercubes. For a vertex $v$ we denote its *antipodal vertex* by $-v$ which is the uniquely determined vertex with maximal distance $N$ from $v$ and which is obtained from $v$ by multiplying all its coordinates by $-1$. Then by (4.2) we have $\varphi_I(-v) = (-1)^{|I|}\varphi_I(v)$ and thus by (4.3) every eigenfunction $f$ corresponding to eigenvalue $2p$ is either *symmetric* or *skewsymmetric*, $f(-v) = (-1)^p f(v)$. Notice that the scalar product of a symmetric function $g$ and a skewsymmetric function $h$ on a hypercube is always 0, i.e., $\langle g, h \rangle = 0$: To see this, consider $V^+ = \{v \in V: v_1 = +1\}$ and $V^- = \{v \in V: v_1 = -1\}$ be a partition of $V$. Then $V^+$ contains all antipodal vertices to $V^-$ and vice versa. Therefore $\langle g, h \rangle = \sum_{v \in V} g(v) h(v) = \sum_{v \in V^+} g(v) h(v) + \sum_{v \in V^-} g(v) h(v) = \sum_{v \in V^+} g(v) h(v) + \sum_{v \in V^+} g(-v) h(-v) = \sum_{v \in V^+} g(v) h(v) - \sum_{v \in V^+} g(v) h(v) = 0$.

**Theorem 4.18.** *An eigenfunction $f$ with eigenvalue $2p$ has at most*

$$\mathfrak{W}(f) \leq \mathfrak{w}^*_{N,p} = \begin{cases} p+1 & \text{if } p=0 \text{ or } p=1, \\ 2\left(1 + \displaystyle\sum_{k=0}^{p/2-1} \binom{N}{2k}\right) & \text{if } p \text{ is even}, \\ 2\left(1 + \displaystyle\sum_{k=0}^{\lfloor p/2 \rfloor -1} \binom{N}{2k+1}\right) & \text{if } p \text{ is odd} \end{cases} \tag{4.7}$$

*weak nodal domains.*

*Proof.* We are given eigenfunctions $f_1, \ldots, f_n$ of the eigenvalues $\lambda_1 < \lambda_2 \leq \cdots \leq \lambda_n$ of the hypercube. Assume first that $f_k$ is a symmetric eigenfunction of the $k$-th eigenvalue. For the proof of this inequality we proceed analogously to Sect. 3.2. However, now we define the subsets $D_1, \ldots, D_m$ such that each $D_i$ consists of the union of one weak nodal domain and the nodal domain that consists of the respective antipodal vertices. First we show that each $D_i$ consists of either one or two nodal domains. Let $D$ be a nodal domain with two adjacent vertices $u$ and $v$. Then $u$ and $v$ differ in exactly one coordinate and $f_k(u)f_k(v) \geq 0$. Moreover, $-u$ and $-v$ also differ in exactly one coordinate and as $f_k(-v) = (-1)^p f_k(v)$ we have $f_k(-u) \cdot f_k(-v) \geq 0$, i.e., the antipodal vertices $-u$ and $-v$ belong to the same nodal domain. Therefore the set $-D$ of all vertices that are antipodal to vertices in $D$ induce a connected subgraph. Hence $D \cup (-D)$ cannot have more than two components. Now define

$$g_i(x) := \begin{cases} f_k(x) & \text{if } x \in D_i, \\ 0 & \text{otherwise}, \end{cases}$$

as in (3.2). Notice that all $g_i$ are symmetric as we have assumed that $f_k$ is symmetric. Thus each $g_i$ is orthogonal to every skewsymmetric eigenfunction $f_j$. Now we can construct a new nonzero function $g := \sum_{i=1}^m a_i g_i$ that is orthogonal to all symmetric eigenfunctions $f_j$, $1 \leq j \leq k-1$. Let $k_s$ be the position of $\lambda_k$ in the subset of eigenvalues with symmetric eigenfunctions. Using the same arguments as in Sect. 3.2 we find that $m$ cannot exceed $k_s$. As every $D_i$ might consist of two nodal domains we have $\mathfrak{W}(f_k) \leq 2k_s$. The analogous result holds when $f_k$ is a skewsymmetric eigenfunction. Thus similarly to (4.6) the proposed inequality follows. □

Inequality (4.7) improves Courant's bound (4.6), see Table 4.1. It seems not to be sharp in general but it gives a sharp bound for eigenvalue $\lambda = 4$ ($p = 2$): $\mathfrak{S}(f) \leq 4$. Notice that this idea cannot be used to improve the upper bound for the number of strong nodal domains.

For the problem of finding the minimum number of nodal domains we will use recursion (4.4) to obtain better bounds. The following technical result will be used.

**Lemma 4.19 ([21]).** *Let $f$ be any function on $K_2^N$ and let $f^+$ and $f^-$ be functions as defined in (4.5). Then $\mathfrak{W}(f^+) = \mathfrak{W}(f)$, $\mathfrak{S}(f^+) = \mathfrak{S}(f)$, $\mathfrak{W}(f^-) \leq 2\mathfrak{W}(f)$, and $\mathfrak{S}(f^-) = 2\mathfrak{S}(f)$.*

**Theorem 4.20 ([21]).** *For all $1 \leq p \leq N-1$ there is an eigenfunction $f$ of the Boolean hypercube with eigenvalue $\lambda = 2p$ such that $\mathfrak{W}(f) = 2$.*

*Proof.* We will proceed by induction using the above technical lemma. The hypercube $K_2^2$ is a cycle with four vertices and has eigenfunction $f_1^{(2)} = \varphi_{\{1\}} + \varphi_{\{2\}}$ of eigenvalue $\lambda = 2p = 2$ which has one positive, one negative and two zero vertices. Thus $\mathfrak{W}(f_1^{(2)}) = 2$. Moreover, both the positive and the negative weak nodal domain contains all vertices where $f_1^{(2)}$ vanishes. Now for $N \geq 3$ we construct eigenfunctions of $K_2^N$ recursively,

$$f_p^{(N+1)} = \binom{1}{1} \otimes f_p^{(N)} = \left(f_p^{(N)}\right)^+ \qquad \text{for } p \leq N-1,$$
$$f_N^{(N+1)} = \binom{1}{-1} \otimes f_{N-1}^{(N)} = \left(f_{N-1}^{(N)}\right)^- \qquad \text{for } p = N,$$

where we use the notation of (4.5). Recall that $f_p^{(N+1)}$ is an eigenfunction of $K_2^{N+1}$ with eigenvalue $2p$. Thus by Lemma 4.19 we find for $p \leq N-1$,

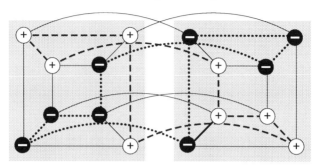

**Fig. 4.4.** Sign pattern of an eigenfunction $f$ on $K_2^4$ with $\mathfrak{S}(f) = 2$. (Figure is taken from [21], © Elsevier 2004)

$\mathfrak{W}(f_p^{(N+1)}) = \mathfrak{W}(f_p^{(N)}) = 2$, where the second equality holds by assumption of induction.

Now consider $f_{N-1}^{(N)}$. By induction we can assume that $\mathfrak{W}(f_{N-1}^{(N)}) = 2$ and hence by Lemma 4.19, $\mathfrak{W}(f_N^{(N+1)}) \leq 4$. Moreover, by construction both the positive and the negative weak nodal domain contains all vertices where $f_{N-1}^{(N)}$ vanishes. The set $Z$ of these vertices induce a subgraph $G[Z]$ that consists of two connected components. It is easy to see that the product $G[Z] \square K_2$ also consists of two connected components and that both of these are contained in all positive and all negative weak nodal domains. Thus there are only two weak nodal domains, i.e., $\mathfrak{W}(f_N^{(N+1)}) = 2$, as proposed. □

For the minimum number of strong nodal domains of eigenfunctions for other eigenvalues the situation is different. Here we only have partial results.

**Theorem 4.21 ([21]).** *For all $1 \leq p \leq \lfloor N/2 \rfloor$ there is an eigenfunction $f$ of the Boolean hypercube with eigenvalue $\lambda = 2p$ such that $\mathfrak{S}(f) = 2$.*

*Remark 4.22.* When we can find a partition $(A, B)$ of the vertex set of $K_2^N$ with $|A| = |B|$ such that the induced subgraphs $G[A]$ and $G[B]$ are connected and $k$-regular, then we can construct an eigenfunction $f$ of eigenvalue $\lambda = 2(N-k)$ with $\mathfrak{S}(f) = 2$ by setting $f(x) = 1$ for all $x \in A$ and $f(y) = -1$ for all $y \in B$. Figure 4.4 shows such an eigenfunction for $K_2^4$ and $k = 2$. We can find such a partition for $k = \lfloor N/2 \rfloor$, see [21, Thm. 4]. Whether such a partition exists for $2 \leq k < \lfloor N/2 \rfloor$ is an open problem. The special case $k = 2$ is of interest in itself: If such a construction exists, there are two disjoint snakes (induced cycles) with length $N/2$.

Numerical experiments described in Table 5.1 in the next chapter (p. 73) show that it should be possible to improve Thm. 4.21:

*Conjecture 4.23.* For all $1 \leq p \leq N - 2$ there is an eigenfunction $f$ of the Boolean hypercube with eigenvalue $\lambda = 2p$ such that $\mathfrak{S}(f) = 2$.

The eigenfunction corresponding to the highest eigenvalue (which is simple) always has $|V|$ (weak or strong) nodal domains by Thm. 4.17, i.e., for the eigenfunction $f$ of the highest eigenvalue we have $\mathfrak{S}(f) = 2^N$. For the second largest eigenvalue there exists a lower bound, which is not sharp (see Table 5.1).

**Theorem 4.24 ([21]).** *For every eigenfunction $f$ of the Boolean hypercube $K_2^N$, $N \geq 3$, with eigenvalue $\lambda = 2(N-1)$ we have $\mathfrak{S}(f) \geq N$.*

*Proof.* Each such eigenfunction $f$ can be expressed as $f(v) = \sum_{j=1}^{N} a_j \phi_j(v)$, where $\phi_j = \varphi_{\{1,\dots,N\}\setminus\{j\}}$, see (4.3). Assume that all coefficients $a_j \geq 0$. Define a new function $\bar{f}$ by $\bar{f}(v) = f(v) \prod_{i=1}^{N} v_i$, where $v_j$ denotes the $j$-th coordinate of vertex $v$, which is either $1$ or $-1$. Then a straightforward computation shows that $\bar{f}(v) = \sum_{j=1}^{N} a_j v_j$. Thus $\bar{f}$ is monotonically decreasing on every path of length $N$ from vertex $\mathbf{1} = (1, \dots, 1)$ to its antipodal $-\mathbf{1}$ and hence $\bar{f}$ changes sign exactly once either on an edge (between two vertices) or on a vertex where $f$ vanishes. (It moreover can be shown that $\bar{f}$ is an eigenfunction of eigenvalue 2.)

Consequently $f$ changes sign exactly $(N-1)$ times on each path from $\mathbf{1}$ to $-\mathbf{1}$. Since every such path is isometric in $K_2^N$, vertices of the same sign that are not adjacent in this path cannot belong to the same nodal domain. Thus such a path intersects exactly $N$ (different) nodal domains and the proposition follows. If some $a_j$ are less than 0 then $\mathbf{1}$ has to be replaced by the vertex with coordinates $(\operatorname{sign}(a_1), \dots, \operatorname{sign}(a_N))$.  $\square$

# 5

# Computational Experiments

It is relatively easy to compute the number of nodal domains for a given eigenfunction[1]. Thus it is no problem to compute the possible number of nodal domains when all eigenvalues are simple. The situation changes completely in the case of degenerate eigenvalues because then the number of nodal domains may vary considerably depending on which eigenfunction from the $r$-dimensional eigenspace of $\lambda_k$ is chosen. Hence, given a fixed graph $G(V, E)$ and an eigenvalue $\lambda_k$ of multiplicity $r$ three questions immediately arise:

- What is the minimal number of nodal domains of $f$?
- What is the maximal number of nodal domains of $f$?
- What is the "typical" number of nodal domains of a corresponding eigenfunction $f$?

In this chapter we deal with the problem of computing possible values for the number of nodal domains of an eigenfunction $f$ for a degenerate eigenvalue and show some results for the class of Boolean hypercubes (see Sect. 4.3.1). Despite the "simplicity" of this class our results are far from being exhausting. The presentation of this chapter mainly follows [21].

## 5.1 Nodal Domains and Hyperplane Arrangements

The eigenspace of an eigenvalue $\lambda$ of multiplicity $r$ can be spanned by a set of orthonormal functions $u_1, \ldots, u_r$. Every eigenfunction $f$ corresponding to eigenvalue $\lambda$ is then given by

---

[1] We assume here that we have algorithms available that compute eigenvectors of symmetric matrices with sufficient accuracy. Of course, round-off errors and deletion of significant digits in floating point arithmetic cause serious problems; especially when an eigenvector (almost) vanishes on one or more components such errors can change its sign pattern. However, this is not topic of this book and we do not discuss this problem in this chapter.

$$f(v) = \sum_{j=1}^{r} a_j \, u_j(v) = \langle \mathbf{a}, \mathbf{u}(v) \rangle$$

where $\mathbf{a} = (a_1, \ldots, a_r)$, and $\mathbf{u}(v) = (u_1(v), \ldots, u_r(v))$ is the vector that contains the values of the basis at the vertex $v$. The convex hull of the vectors $\mathbf{u}(v)$, for $v \in V$, forms a polytope in $\mathbb{R}^r$, which is called the *eigenpolytope* of the graph, see e.g. [28, 84].

It is obvious that the number of (strong or weak) nodal domains only depends on the signs of the eigenfunction on each vertex. There is a one-to-one relation between the eigenfunction $f$ and its "coordinate vector" $\mathbf{a}$. The sign at vertex $v$ is given by the sign of $\langle \mathbf{a}, \mathbf{u}(v) \rangle$ and thus the set of eigenfunctions that vanish on vertex $v$ corresponds to the set

$$H_v = \{ a \in \mathbb{R}^r : \langle \mathbf{a}, \mathbf{u}(v) \rangle = 0 \}$$

which is either a hyperplane through the origin in $\mathbb{R}^r$ or, if $\mathbf{u}(v) = 0$, $H_v = \mathbb{R}^r$. The set of all proper hyperplanes forms a *hyperplane arrangement*

$$\mathcal{H} = \{ H_v : v \in V \}$$

in $\mathbb{R}^r$, see e.g. [60, 180]. The union of all these hyperplanes creates a *cellular complex* in $\mathbb{R}^r$ or (if we look at normalized eigenfunctions) in the sphere $\mathbb{S}^{r-1}$. A cellular complex consists of disjoint cells, where each cell is either homeomorphic to an open disc $D_d = \{ \mathbf{x} \in \mathbb{R}^d : ||\mathbf{x}||_2 < 1 \}$ or a single point. In the former case we say that the cell has dimension $d$ and the cell is called a *d-cell*. In the latter case we have a *0-cell*. Additionally, a cellular complex satisfies the following properties: (i) the union of all cells is the entire space $\mathbb{R}^r$ (or $\mathbb{S}^{r-1}$); (ii) the boundary of a $d$-cell consists of the union of cells of dimension less than $d$.

Each of the hyperplanes $H_v$ splits the $\mathbb{R}^r$ into three pieces: the hyperplane $H_v$ itself and the two open half-spaces $\{ \mathbf{a} \in \mathbb{R}^r : \langle \mathbf{a}, \mathbf{u}(v) \rangle > 0 \}$ and $\{ \mathbf{a} \in \mathbb{R}^r : \langle \mathbf{a}, \mathbf{u}(v) \rangle < 0 \}$. Hence, for each vector $\mathbf{a} \in \mathbb{R}^r$ we may introduce the *covector* or *position vector* $\mathbf{c_a}$ which has $\text{sign}\langle \mathbf{a}, \mathbf{u}(v) \rangle$ as its components. (Using our notion then the covector $\mathbf{c_a}$ is a function on $V$.) Obviously the covector $\mathbf{c_a}$ is constant in each cell of the cellular complex and it uniquely determines each cell. Moreover, it corresponds to the sign pattern of the associated eigenfunction. Consequently, finding all possible values for the number of nodal domains is equivalent to finding all cells of this complex. However, the number of cells explodes with the number of vertices and the multiplicity $r$ of the eigenvalue. Using a general upper bound for hyperplane arrangements [60] we have the asymptotic behavior

$$\text{number of } d\text{-cells} \sim |V(G)|^r. \tag{5.1}$$

An exact and sharp upper bound is given, e.g., in [60].

The following observations will simplify our task. Assume that we go along a path within a cell towards its boundary. As long as we stay inside the cell

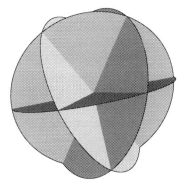 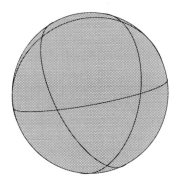

**Fig. 5.1.** Hyperplane arrangement (l.h.s.) and the corresponding cells on the sphere (r.h.s.) of eigenvalue 4 for the hypercube $K_2^3$. We have $r = 3$ and $|V(G)| = 8$. The vectors $\mathbf{u}(v)$ are given by the eight vectors $(\pm 1, \pm 1, \pm 1)$. Due to symmetry we only have the following cells

| dim | shape | $\mathfrak{S}(f)$ | $\mathfrak{W}(f)$ |
|---|---|---|---|
| 2 | rectangle | 4 | 4 |
| 2 | triangle | 3 | 3 |
| 1 | edge | 4 | 3 |
| 0 | point | 3 | 2 |

on the sphere $\mathbb{S}^2$. This is easily checked using `Mathematica`. (Figure is taken from [21], © Elsevier 2004)

nothing happens and the number of nodal domains remains unchanged. But if we reach the boundary the eigenfunction vanishes on some (but at least one) of the nonzero vertices whereas all other remain unchanged. This has two consequences:

The number of weak nodal domains is either decreasing or remains constant, since zero vertices do not separate weak nodal domains. So we have to look at 0-cells if we want to minimize $\mathfrak{W}(f)$ and to cells of highest dimension if we want to maximize $\mathfrak{W}(f)$, for the eigenfunction $f$ of $\lambda$.

The situation is much more complicated at strong nodal domains, because then zeros separate nodal domains, and $\mathfrak{S}(f)$ may increase. However, if the eigenfunction vanishes on too many vertices when we reach the boundary, it might happen that nodal domains disappear which decreases $\mathfrak{S}(f)$. This happens for example with some eigenfunctions of the second eigenvalue of stars (connected graphs where all but one vertex have degree 1), or more generally with some eigenfunctions corresponding to eigenvalues where Faria vectors exist. Figure 5.1 illustrates the situation.

## 5.2 A Hillclimbing Algorithm

Because of (5.1) it is in practice impossible to calculate all cells of a hyperplane arrangement for any reasonably sized graph. We have therefore devised a *hillclimbing algorithm* to search for the minimum (or maximum) number of strong (or weak) nodal domains. This algorithm is based on the above observations, moving from a cell to neighboring cells in search of an improved number of nodal domains.

Briefly, the algorithm works as follows. Starting from some random point $\mathbf{a}$ in the hyperplane arrangement with corresponding eigenfunction $f(v) = \langle \mathbf{a}, \mathbf{u}(v) \rangle$. Pick a second random point $\mathbf{a}'$ and move into the direction of this second point until a boundary in the cellular complex is crossed, i.e., at least one of the coordinates of the position vector has changed sign and a neighboring cell is entered. To this end we define $\delta(v) = \langle \mathbf{a}, \mathbf{u}(v) \rangle / \langle \mathbf{a}', \mathbf{u}(v) \rangle$, and find the vertices $v_1$ and $v_2$ such that $\delta(v_1)$ is smallest with $\delta(v) > 0$ and $\delta(v_2)$ is smallest with $\delta(v) > \delta(v_1)$. Then set $\delta = (\delta(v_1) + \delta(v_2))/2$ and move from $\mathbf{a}$ to $\mathbf{a}^* = \mathbf{a} - \delta \mathbf{a}'$, with corresponding eigenfunction $f'(v) = \langle \mathbf{a}^*, \mathbf{u}(v) \rangle$. If the number of (strong) nodal domains of this new cell is less than or equal to that of the cell that was moved from, accept this move, i.e., make the new point the current one. Otherwise, return to the original point, i.e., do not update the current point. (If we want to find the maximum number of nodal domains we proceed when the number of (strong) nodal domains of this new cell is greater than or equal to that of the cell that was moved from.) Now repeat this sequence of picking a random second point, moving towards it from the current point until a cellular boundary is crossed, and determining whether the move is accepted or not, until some stopping criterion is reached.

Notice that the algorithm also accepts neutral moves, i.e., moves to neighboring cells that have an equal number of nodal domains. This way, getting stuck in the middle of some plateau is avoided. Since it is not obvious with this "random move" algorithm when a local optimum is reached, we terminate the search when the number $R$ of moves without improvement exceeds a user-defined upper bound.

It must be noted here that this algorithm only deals with coordinate vectors in cells of highest dimension correctly, i.e., the corresponding eigenfunctions have no vanishing vertices (except those vertices where all eigenfunctions of the given eigenvalue vanish). It can be adopted such that it also includes searching on cells of lower dimension. However, there are some difficult numerical problems that require sophisticated methods from computational geometry.

## 5.3 Numerical Experiments for the Boolean Hypercube

In the following we apply the above method to get an idea of possible numbers of nodal domains for eigenfunctions of the Laplacian of the Boolean hypercube. We start with an investigation of the "typical number of nodal domains". For this task we first must be precise about the meaning of this term. We have seen in Sect. 4.3.1, Eq. (4.3), that every eigenfunction in the eigenspace $\mathcal{E}_\lambda = \{ f : \mathbf{L}f = \lambda f \}$ of eigenvalue $\lambda = 2p$ can be represented as

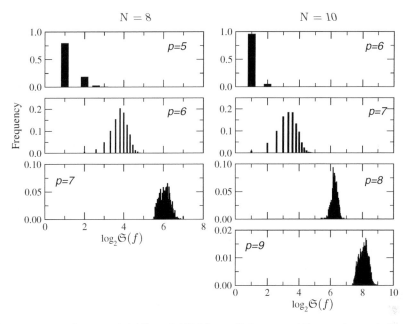

**Fig. 5.2.** Distribution of $\mathfrak{S}(f)$ with Walsh coefficients $a_I$, $|I| = p$, drawn independently from a Gaussian distribution. (Results from [21])

$$f(v) = \sum_{I\,:\,|I|=p} a_I\,\varphi_I(v) \,. \tag{5.2}$$

Thus the set of the Walsh functions $\varphi_I$ with $|I| = p$ which forms an orthonormal basis of $\mathcal{E}_{2p}$. Consequently, in order to define more precisely what we mean by the "typical number of nodal domains" we have to specify a distribution of the coefficients $a_I$.

From a physics point of view it is most natural to assume that the $a_I$ are independent identically distributed Gaussian random variables. In this case (5.2) defines Derrida's $p$-spin models [51, 52] that form an important and well-studied class of spin glasses which also play an important rôle in the theory of fitness landscapes [160].

If we use the hyperplane arrangement described above we might be interested in the volume of the cells that correspond to a given number of nodal domains. This volume is very hard to compute, but it can be done approximately using Monte Carlo integration (see e.g. [71]). For this purpose the coefficient vectors are sampled from a uniform distribution on the corresponding sphere.

Fortunately these two pictures are equivalent. Normalizing random vectors that follow a multivariate Gaussian law (as in the first approach) gives uniformly distributed points on the sphere (see e.g. [54, 99]). The empirical distribution of the number of strong nodal domains $\mathfrak{S}(f)$ for $K_2^8$ and $K_2^{10}$

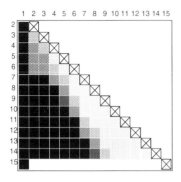
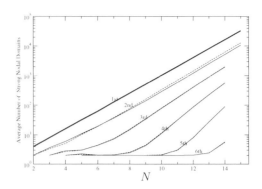

**Fig. 5.3.** Average number of strong nodal domains for the eigenfunctions of the hypercubes with $N = 2$ to 15 as a function of $p$. The l.h.s. panel gives an overview of the numerical survey. Black squares denote $(N, p)$-pairs for which all of the 1000 randomly generated instances had exactly 2 nodal domains, $\boxtimes$ denotes the $2^N$ nodal domains for $p = N$, and the gray boxes denote average numbers of strong nodal domains in the ranges 2–3, 4–10, and greater than 10.
The r.h.s. panel displays the average number of strong nodal domains of the $k$-th largest eigenfunctions as a function of $N$ (here we do not count multiplicities). Note that the largest eigenvalue is unique and has the maximally possible number of $|V| = 2^N$ strong nodal domains. (Results from [21])

and some eigenvalues obtained by such a Monte Carlo integration is shown in Fig. 5.2. Figure 5.3 displays the average numbers of nodal domains for all pairs $(N, p)$ of dimensions $N$ (up to $N = 15$) and eigenvalues $2p$.

For the maximum number of weak nodal domains and the minimum number of strong nodal domains the hill-climbing algorithm as described in Sect. 5.1 gives some results – shown in Table 5.1 – that let us conjecture that the bounds from Sect. 4.3.1 are not sharp in general.

We suspect that the bounds in Table 5.1 for the minimum number of strong nodal domains for the 2nd largest eigenvalue are sharp at least for $N \leq 10$. However, the sequence 2, 3, 4, 8, 14, 24, 44, 84, 160, ... does not appear to be a known integer sequence.

*Remark 5.1.* For reasons that we do not fully understand maximizing the number of nodal domains on a given eigenspace seems to be much harder than minimizing. This difference in difficulty between minimizing and maximizing the number of nodal domains deserves an explanation. One explanation could be that the hill-climbing algorithm often runs into local minima which are also global minima; whereas for maximization there are many local but not global maxima where the algorithm gets stuck.

*Remark 5.2.* A direct computational approach for the maximum number of strong nodal domains fails because we would have to compute all cells of dimension 0; this is not only numerically difficult but the number of 0-cells is also too large. A completely different approach is therefore required.

**Table 5.1.** Upper and lower bounds on the number of nodal domains as functions of $N$ and $p$ found by numerical experiments using a hill-climbing algorithm. (Results from [21])

| $p =$ | 1 | 2 | 3 | 4 | 5 | 6 | 7 | 8 | 9 | 10 | 11 | 12 | 13 | 14 |
|---|---|---|---|---|---|---|---|---|---|---|---|---|---|---|
| $N$ | Upper Bounds on Minimal Number of Strong Nodal Domain ||||||||||||||

| $N$ | 1 | 2 | 3 | 4 | 5 | 6 | 7 | 8 | 9 | 10 | 11 | 12 | 13 | 14 |
|---|---|---|---|---|---|---|---|---|---|---|---|---|---|---|
| 2 | 2 | 4 | | | | | | | | | | | | |
| 3 | 2 | 3 | 8 | | | | | | | | | | | |
| 4 | 2 | 2 | 4 | 16 | | | | | | | | | | |
| 5 | 2 | 2 | 2 | 8 | 32 | | | | | | | | | |
| 6 | 2 | 2 | 2 | 2 | 14 | 64 | | | | | | | | |
| 7 | 2 | 2 | 2 | 2 | 2 | 24 | 128 | | | | | | | |
| 8 | 2 | 2 | 2 | 2 | 2 | 2 | 44 | 256 | | | | | | |
| 9 | 2 | 2 | 2 | 2 | 2 | 2 | 2 | 84 | 512 | | | | | |
| 10 | 2 | 2 | 2 | 2 | 2 | 2 | 2 | 2 | 160 | 1024 | | | | |
| 11 | 2 | 2 | 2 | 2 | 2 | 2 | 2 | 2 | 2 | 314 | 2048 | | | |
| 12 | 2 | 2 | 2 | 2 | 2 | 2 | 2 | 2 | 2 | 2 | 620 | 4096 | | |
| 13 | 2 | | | | | | | | | | | 1280 | 8192 | |
| 14 | 2 | | | | | | | | | | | | 2446 | 16384 |

$N$ — Lower Bounds on Maximal Number of Weak Nodal Domain[†]

| $N$ | 1 | 2 | 3 | 4 | 5 | 6 | 7 | 8 | 9 |
|---|---|---|---|---|---|---|---|---|---|
| 2 | 2 | 4 | | | | | | | |
| 3 | 2 | 4 | 8 | | | | | | |
| 4 | 2 | 4 | 8 | 16 | | | | | |
| 5 | 2 | 4 | **10** | 16 | 32 | | | | |
| 6 | 2 | 4 | 8 | **18** | 32 | 64 | | | |
| 7 | 2 | 4 | *4* | *15* | **34** | 64 | 128 | | |
| 8 | 2 | *2* | | | *12* | *57* | 128 | 256 | |
| 9 | 2 | | | | | *72* | **261** | 512 | |

[†] Numbers in **bold** are bounds that are better then Thm. 4.17. Moreover, they (again) show that one cannot deduce an upper bound for the number of weak or strong nodal domains of a product graph when only bounds of each of its factors are known. Entries in *italics* are numerical value that are known to be underestimates because of Lemma 4.19.

## 5.4 Local Optima

The number of nodal domains of a function $f$ reflects its "ruggedness". In the introduction (Sect. 1.4) we briefly discussed correlation measures of ruggedness. Another characterization of ruggedness is the number $n_{\text{loc.opt.}}$ of local optima [142], or more precisely the number

$$\eta = \ln n_{\text{loc.opt.}} / \ln n . \tag{5.3}$$

Very little is known analytically about the number and distribution of local optima. Numerical data are available for a few combinatorial optimization problems obtained from sampling random elements $x \in V$ and checking whether they are local optima [79, 162, 163]. Methods from statistical mechanics can be used to obtain the asymptotic behavior of $\eta$ for a family of landscapes on graphs of increasing size [25, 50, 53, 63, 89, 150, 166, 167]. These approaches do not make use of the fact that $f$ is an eigenfunction of a Laplacian but rather they are based upon a representation of $f$ as a linear combination and knowledge on the distribution of the coefficients of this expansion.

On the Boolean hypercube we observe that eigenfunctions with eigenvalue $\lambda_2 = 2$, which are of the form $f(x) = \sum_{i=1}^{N} a_{\{i\}} \varphi_{\{i\}}(x)$, generically have a single local minimum define by $x_i = -\text{sign}\, a_i$ for all $i$. This is not true, however, on all graphs. At present, it is an open problem whether there is close connection between local optima of a function $f$ on a graph and spectral properties.

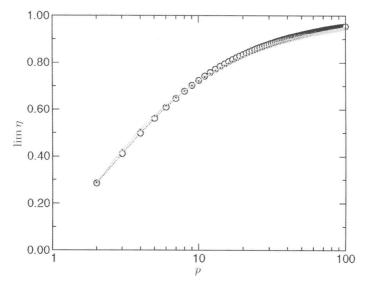

**Fig. 5.4.** Asymptotic density $\eta$ of local minima in typical $p$-spin landscapes, i.e., eigenfunction of the eigenvalue to the eigenvalue $\lambda = 2p$ with coefficients $a_I$ randomly chosen on the sphere. Full line: statistical mechanics solution based on the so-called TAP equations [89], open circles: estimates from the correlation length conjecture. Data are from [158].

We mention a nonrigorous result here, because it approximates $\eta$ surprisingly well in "typical" elementary landscapes landscapes, namely those which are of the form $f = \sum_j a_j f_j$ where the $f_j$ form an orthonormal basis of an eigenspace $\mathcal{E}_\lambda$ of the graph Laplacian and the coefficients $a_j$ are independent, identically distributed Gaussian random variables. Such models are termed isotropic in [160]. We remark that these models maximize entropy given the autocorrelation function $r(s)$ [158]. In [163] it has been suggests that the number of local optima of such a "typical" landscape can be estimated from its correlation length $\ell$, Eq. (1.8). More precisely, one expects only a small number of local optima within a ball $B$ in $G$ with a radius $R = R(\ell)$ that is given by the expected distance between two vertices separated by random walk of length $\ell$ on $G$. This simple idea yields very good estimates for $\eta$ for a wide variety of models [79, 162]. Figure 5.4 shows the typical eigenfunction of the Boolean hypercube as an example.

# 6

# Faber-Krahn Type Inequalities

The celebrated *Faber-Krahn Theorem* gives an important isoperimetric inequality concerning Dirichlet eigenvalues. It states that the ball has lowest first Dirichlet eigenvalue amongst all bounded domains of the same volume in $\mathbb{R}^s$ (with the standard Euclidean metric). It has been first conjectured by Rayleigh and proved independently by Faber [61] and Krahn [118] for the $\mathbb{R}^2$; a proof of the generalized version can be found for example in [29]. The Faber-Krahn theorem can also be rephrased in the following way: for all drums with the same area and same tension the circular-shaped has the lowest tone.

Discrete Faber-Krahn theorems are of course concerned primarily with the geometry of domains rather then eigenfunctions. They naturally fit into the framework of this monograph because their proofs heavily rely on explicit constructions for the eigenfunction of the first Dirichlet eigenvalue. In addition we discuss some properties of Dirichlet eigenfunctions in general.

In Section 1.5 we have introduced the notion of graphs with boundaries and discrete Dirichlet operators and thus we should be able to formulate similar inequalities. We additionally have to introduce the notion of the *volume of a graph*. However, it is not immediately clear how to transform the continuous result to the discrete setting of graphs. We derive two approaches. The first one is due to Friedman [75] who imitates the situation of the continuous case: Take the $\mathbb{R}^2$ (an "infinitely large skin"), take a pair of scissors and cut out a piece for the drum. In the discrete setting we look at the geometric realization of an infinite regular tree and cut out a subtree. Wherever we cut an arc we insert a boundary vertex and fix the "membrane" (see Fig. 6.4 on p. 84). For the volume we used the total length of the edges of these subtrees (Sect. 6.4). For the second approach we look at more general classes of (not necessarily regular) trees where the volume is the number of (interior) vertices (Sect. 6.3). The main ideas of the proofs are given in Sects. 6.5 and 6.6.

## 6.1 Basic Properties of Dirichlet Operators

In Section 1.5 we have defined the *discrete Dirichlet operator*. In Friedman's setting this definition has to be extended to weighted graphs as it has been done in Sect. 2.2 for the graph Laplacian. Although this is completely analogous we repeat the necessary definitions for the sake of a more compact presentation.

Let $G(V^\circ \cup \partial V, E^\circ \cup \partial E)$ be a (weighted) connected graph with boundary and $\mathbf{L}_w$ the (weighted) graph Laplacian (see Sect. 2.2). Remind that there are no edges between boundary vertices. Unweighted graphs can be seen as special case of graphs with weights $w_{xy} = 1$ if $x$ and $y$ are adjacent and $w_{xy} = 0$ otherwise. A solution $f$ to the Dirichlet problem must satisfy $\mathbf{L}_w f(v) = \lambda f(v)$ for all interior vertices $v \in V^\circ$ and $f(z) = 0$ for all boundary vertices $z \in \partial V$. Equivalently, we can compute the eigenvalues and eigenfunctions of the corresponding discrete Dirichlet operator $\mathbf{L}_w^\circ(G)$ which is derived from the graph Laplacian $\mathbf{L}_w(G)$ simply by deleting all rows and columns that correspond to boundary vertices. We call the eigenvalues of this operator the *Dirichlet eigenvalues* of the graph with boundary. We are interested in the first Dirichlet eigenvalue which we denote by $\lambda^\circ(G)$. It can be computed by

$$\lambda^\circ(G) = \min_{f \in \mathcal{F}_0} \mathcal{R}_{\mathbf{L}_w}(f) = \min_{f \in \mathcal{F}_0} \frac{\sum_{xy \in E} w_{xy}(f(x) - f(y))^2}{\sum_{x \in V} f(x)^2} \tag{6.1}$$

where $\mathcal{F}_0$ denotes the set of real-valued functions on $V$ with the constraint $f|_{\partial V} = 0$ (Corollary 2.5). Moreover, every function $f \in \mathcal{F}_0$ that minimizes the Rayleigh quotient is an eigenfunction to the first Dirichlet eigenvalue $\lambda^\circ(G)$. The first Dirichlet eigenvalue has the following basic properties that follow immediately from (6.1) and the Perron-Frobenius Theorem (Thm. 2.22 and Cor. 2.23).

**Lemma 6.1 ([75]).** *Let $G(V^\circ \cup \partial V, E^\circ \cup \partial E)$ be a connected graph with boundary. Then*

*(1) $\mathbf{L}_w^\circ(G)$ is a positive operator, i.e. $\lambda^\circ(G) > 0$.*
*(2) An eigenfunction $f$ corresponding to eigenvalue $\lambda^\circ(G)$ is either positive or negative on all interior vertices of $G$.*
    *Thus we assume without loss of generality $f(v) > 0$ for all $v \in V^\circ$.*
*(3) $\lambda^\circ(G)$ is a simple eigenvalue.*

Dirichlet eigenvalues can be viewed as weighted averages of the number of boundary vertices to which interior vertices are connected:

**Theorem 6.2 ([20]).** *Let $G(V^\circ \cup \partial V, E^\circ \cup \partial E)$ be a connected graph with boundary and $f$ an eigenfunction corresponding to some eigenvalue $\lambda$ of the Dirichlet operator. Let $b(v) = \sum_{u \in \partial V} w_{vu}$. (For unweighted graphs $b(v)$ just denotes the number of boundary vertices adjacent to $v$.) Then either $\sum_{v \in V} f(v) = \sum_{v \in V} b(v) f(v) = 0$, or*

$$\lambda = \frac{\sum_{v \in V} b(v) \, f(v)}{\sum_{v \in V} f(v)} \, .$$

*Proof.* Let $\mathbf{1} = (1, \ldots, 1)^\mathsf{T}$, $i(v) = \sum_{u \in V^\circ} w_{vu}$, and $\delta(v) = \sum_{u \in V} w_{vu}$. (For unweighted graphs $i(v)$ denotes the number of interior vertices adjacent to $v$ and $\delta(v) = d(v)$.) Thus $b(v) + i(v) = \delta(v)$. A straightforward computation gives

$$\begin{aligned}
\langle \mathbf{1}, \mathbf{L}_w^\circ f \rangle &= \sum_{v \in V^\circ} \sum_{u \in V} w_{vu} \, (f(v) - f(u)) \\
&= \sum_{v \in V^\circ} \delta(v) \, f(v) - \sum_{v \in V^\circ} \sum_{u \in V} w_{vu} \, f(u) \\
&= \sum_{v \in V^\circ} \delta(v) \, f(v) - \sum_{u \in V} f(u) \sum_{v \in V^\circ} w_{vu} \\
&= \sum_{v \in V^\circ} \delta(v) \, f(v) - \sum_{u \in V^\circ} i(u) \, f(u) \\
&= \sum_{v \in V_0} b(v) \, f(v) \, .
\end{aligned}$$

Since $f$ is an eigenfunction we find $\langle \mathbf{1}, \mathbf{L}_w^\circ f \rangle = \lambda \sum_{v \in V_0} f(v)$. Since $f(v) = 0$ for all boundary vertices $v \in \partial V$ the result follows. □

This result is similar to Thm. 2.1. Indeed we can view $b(v)$ as the nonnegative potential $p(v)$ in (2.5). On the other hand we can replace any positive potential by adding a new boundary edge of weight $p(v)$ to each vertex $v$ of the graph. Then the discrete Dirichlet operator of the new graph coincides with the Laplacian of the original graph. As an immediate consequence the considerations of Remark 2.2 also holds for Thm. 6.2: The case $\sum_{v \in V} f(v) = 0$ happens, for example, for all eigenfunctions of an eigenvalue $\lambda > \lambda_1$ when $b(v)$ is constant for all $v \in V^\circ$.

## 6.2 The Faber-Krahn Property

We say that a graph with boundary has the *Faber-Krahn property* if it has lowest first Dirichlet eigenvalue among all graphs with the same "volume" in a particular graph class. This definition raises two questions:

(1) What is an appropriate graph class, and
(2) What is the "volume" of a graph?

Making the graph class too large leads to quite simple (noninteresting) graphs. For example, if we look at the set of all connected graphs with a given number of edges as the "volume" of the graph and nonempty set of boundary edges, then graphs with the Faber-Krahn property are paths. To be precise the following theorem holds.

**Theorem 6.3 ([110]).** *Let $G(V^\circ \cup \partial V, E^\circ \cup \partial E)$ be a connected graph with (nonempty) boundary where every interior vertex has degree at least 2. If $m = |E^\circ \cup \partial E|$ denotes the number of its edges, then*

$$\lambda^\circ(G) \geq \lambda^\circ(L_m')$$

where $L'_m$ is a path with a terminating triangle and a terminating boundary vertex and m edges as shown in Fig. 6.1(i).

If additionally the "nonseparation property" holds, i.e., each component of $G - v$ contains a boundary vertex, then $\lambda^\circ(G) \geq \lambda^\circ(L''_m)$ where $L''_m$ is the path with $m - 1$ interior vertices and two boundary vertices (and thus m edges) as shown in Fig. 6.1(ii).

Equality holds if and only if G equals the respective graph $L'_m$ or $L''_m$.

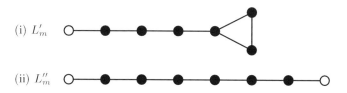

(i) $L'_m$

(ii) $L''_m$

**Fig. 6.1.** The graphs $L'_m$ (i) and $L''_m$ (ii) with $m = 7$ edges.
(● ... interior vertices, ○ ... boundary vertices)

If we drop the restriction that interior vertices have smallest degree at least 2, then the resulting graph is a simple path with only one boundary vertex [109]. If we use the number n of vertices as a measure for the volume of a graph than we have a similar result (this follows from Thm. 6.9(i) below and the fact that the Rayleigh quotient of a graph cannot be smaller than the Rayleigh quotient of any of its spanning trees).

At the time of writing this book more detailed results are known for trees only. While their structure is simple enough to formulate (and prove) Faber-Krahn-type theorems the results are quite surprising for such simple graphs. The partitioning of a graph into boundary and interior vertices is arbitrary as we have seen on p. 9. In the case of trees, however, it seems natural to define the leaves as boundary, i.e., a vertex is a boundary vertex if and only if it has degree 1.

Recall that the geometric realization $\mathcal{G}$ of a graph G is the metric space consisting of V and arcs of length $1/w_{uv}$ glued between u and v for every edge $e = uv \in E$ (see Sect. 2.4).

Motivated by the results for $\mathbb{R}^s$ we would expect that a "ball" will minimize the first Dirichlet eigenvalue. A *ball* $B(c, r)$ is a graph G with boundary with a center $c \in \mathcal{G}$, not necessarily a vertex, and a radius $r > 0$, such that $\text{dist}(c, v) \leq r$ for all points $v \in B(c, r)$, where equality holds if and only if $v \in \partial V$. $\text{dist}(u, v)$ denotes the geodesic distance between $u, v \in \mathcal{G}$.

In the next two sections we state the Faber-Krahn type theorem. From here on we will drop the subscript w in $\mathbf{L}^\circ_w(G)$ and $\mathbf{L}_w(G)$ for weighted graphs G since there is no risk of confusion.

## 6.3 Unweighted Trees

We first consider the case of unweighted trees (where all edges have length 1). It will be convenient to root the trees at a vertex $v_0$. The *height* $h(v)$ of a vertex $v$ in a tree $G$ with root $v_0$ is defined as the geodesic distance of $v$ from the root: $h(v) = \mathrm{dist}(v, v_0)$. For two adjacent vertices $v$ and $w$ with $h(w) = h(v) + 1$ we call $v$ the *parent* of $w$, and $w$ a *child* of $v$. Notice that every vertex $v \neq v_0$ has exactly one parent, and every interior vertex $w$ has at least one child vertex.

If we only fix the total number of vertices and use the number of vertices as measure of the volume of $G$, then the tree that minimizes the first Dirichlet eigenvalue is a path [110]. Hence we will consider more restricted classes of graphs with boundaries in the following. We define

$$\mathcal{T}^{(n,k)} = \{G \text{ is a tree, with } |V| = n \text{ and } |V^\circ| = k\} \,,$$

$$\mathcal{T}_d^{(n,k)} = \{G \in \mathcal{T}^{(n,k)} : d(v) \geq d \text{ for all } v \in V^\circ\} \,.$$

As it is clear that we always look at a particular class $\mathcal{T}^{(n,k)}$ or $\mathcal{T}_d^{(n,k)}$ we will write $\mathcal{T}$ and $\mathcal{T}_d$ for short; $n$ and $k$ have then to be selected accordingly. We always assume that $1 \leq k \leq n - 1$.

Another interesting class is based on so called degree sequences. A sequence $\pi = (d_0, \ldots, d_{n-1})$ of nonnegative integers is called *degree sequence* if there exists a graph $G$ with $n$ vertices for which $d_0, \ldots, d_{n-1}$ are the degrees of its vertices. For trees the following characterization exists.

**Proposition 6.4 ([94]).** *A degree sequence* $\pi = (d_0, \ldots, d_{n-1})$ *is a tree sequence (i.e. a degree sequence of some tree) if and only if every* $d_i > 0$ *and* $\sum_{i=0}^{n-1} d_i = 2(n-1)$.

Using this notion we can introduce the class

$$\mathcal{T}_\pi = \{G \text{ is a tree with boundary with degree sequence } \pi\} \,.$$

Notice that for a particular degree sequence $\pi$ we have

$$\mathcal{T}_\pi \subseteq \mathcal{T}_{d_\pi} \subseteq \mathcal{T}_2 = \mathcal{T}$$

where $d_\pi$ is the minimal degree for interior vertices of the degree sequence $\pi$.

For the class $\mathcal{T}$ of all trees we again find only a simple structure for graphs with the Faber-Krahn property.

**Theorem 6.5 (Klobürşteltheorem[1], [20]).** *A tree $G$ has the Faber-Krahn property in the class $\mathcal{T}$ if and only if $G$ is a star with a long tail, i.e. a comet, see Fig. 6.2. $G$ is then uniquely determined up to isomorphism.*

---

[1] *Klobürştel* is the Viennese term for a comet-shaped cleaning device commonly referred as a toilet brush. The slightly nonstandard spelling stems from the fact that the result was conceived at a meeting in Slovenia.

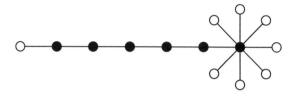

**Fig. 6.2.** A comet has the Faber-Krahn property in class $\mathcal{T}$. It consists of a star with diameter 2 and a path attached to it.
(● ... interior vertices, ○ ... boundary vertices)

Graphs with the Faber-Krahn property in $\mathcal{T}_d$ or $\mathcal{T}_\pi$ have a richer structure. The main notion for describing trees in $\mathcal{T}_d$ or $\mathcal{T}_\pi$ with the Faber-Krahn property is spiral-like ordering of its vertices introduced by Pruss [148]. We give a slightly modified and extended definition.

A well-ordering $\prec$ of the vertices is called *spiral-like* (*SLO-ordering* for short) if the following holds for all vertices $v, v_1, v_2, w, w_1, w_2 \in V$:

(S1) $v \prec w$ implies $h(v) \leq h(w)$, where $h(v)$ denotes the height of $v$;
(S2) if $v_1 \prec v_2$ then for all children $w_1$ of $v_1$ and all children $w_2$ of $v_2$, $w_1 \prec w_2$;
(S3) if $v \prec w$ and $v \in \partial V$, then $w \in \partial V$.

It is called *spiral-like with increasing degrees* (*SLO\*-ordering* for short) if additionally the following holds

(S4) if $v \prec w$ for interior vertices $v, w \in V^\circ$, then $d(v) \leq d(w)$.

We call trees that have a SLO- or SLO\*-ordering of its vertices *SLO-trees* and *SLO\*-trees*, respectively.

Notice that SLO-trees are almost balls, i.e., there exists a vertex $c$ and a radius $r$ such that $B(c, r) \subseteq G \subset B(c, r+1)$, see Fig. 6.3 for an example.

**Theorem 6.6 ([20]).** *A graph $G$ has the Faber-Krahn property in a class $\mathcal{T}_d$ if and only if it is a SLO\*-tree where at most one interior vertex has degree $d^\circ$ exceeding $d$ and all other interior vertices have degree $d$. $G$ is then uniquely determined up to isomorphism.*

**Theorem 6.7 ([20]).** *A graph $G$ with degree sequence $\pi$ has the Faber-Krahn property in the class $\mathcal{T}_\pi$ if and only if it is a SLO\*-tree. $G$ is then uniquely determined up to isomorphism.*

The main ideas of the proofs of Thms. 6.6 and 6.7 are outlined in Sects. 6.5 and 6.6. As an immediate corollary we get a result of Pruss [148].

**Corollary 6.8 ([148, Thm. 6.2]).** *In the class of d-regular unweighted trees a graph $G$ has the Faber-Krahn property if and only if it is a SLO\*-tree. $G$ is then uniquely determined up to isomorphism.*

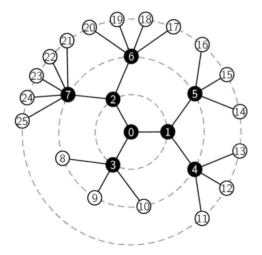

**Fig. 6.3.** A SLO*-tree with 8 interior and 18 boundary vertices. The SLO*-ordering $\prec$ is indicated by numbers. Degree sequence $\pi = (3, 3, 3, 4, 4, 4, 5, 6, 1, 1, \ldots, 1)$.

One might ask what happens when the conditions defining $\mathcal{T}^{(n,k)}$ and $\mathcal{T}_d^{(n,k)}$ are relaxed. It seems natural to consider the classes

$$\mathcal{T}^{(n,\cdot)} = \{G \text{ is a tree, with } |V| = n\}$$

$$\mathcal{T}_d^{(n,\cdot)} \text{ and } = \{G \in \mathcal{T}^{(n,\cdot)} : d_v \geq d \text{ for all } v \in V^\circ\},$$

where we keep the total number of vertices fixed, and

$$\mathcal{T}^{(\cdot,k)} = \{G \text{ is a tree, with } |V^\circ| = k\} \text{ and}$$

$$\mathcal{T}_d^{(\cdot,k)} = \{G \in \mathcal{T}^{(\cdot,k)} : d_v \geq d \text{ for all } v \in V^\circ\},$$

where we keep the number of interior vertices fixed. Using the same ideas as in the proofs of Thms. 6.6 and 6.7 we find the following characterizations for graphs with the Faber-Krahn property:

**Theorem 6.9 ([20]).** *A tree $G$ with boundary has the Faber-Krahn property*

(i) *in $\mathcal{T}^{(n,\cdot)}$ if and only if it is a path with $n$ vertices.*

(ii) *in $\mathcal{T}_d^{(n,\cdot)}$ if and only if it is a SLO*-tree where exactly one interior vertex has degree $d^\circ$ with $d \leq d^\circ < 2\,d$ and all other interior vertices have degree $d$. (This is the SLO*-tree in $\mathcal{T}_d^{(n,\cdot)}$ with the greatest number of interior vertices.)*

(iii) *in $\mathcal{T}^{(\cdot,k)}$ if and only if it is a path with $k + 2$ vertices.*

(iv) *in $\mathcal{T}_d^{(\cdot,k)}$ if and only if it is a SLO*-tree where all interior vertices have degree $d$.*

*$G$ is then uniquely determined up to isomorphism.*

## 6.4 Semiregular Trees

Friedman [75] considers the class of trees that consist of (connected) subsets of the geometric realization of the $d$-regular infinite tree. These graphs are trees where all interior vertices have degree $d$, all interior edges have length (weight) 1 and all boundary edges have length at most 1, see Fig. 6.4. The volume is then the total length of all edges, i.e., the measure $\mu_E$ defined in Sect. 2.4. Notice that this is equivalent to the number of vertices for unweighted trees. We call such trees $d$-semiregular trees.

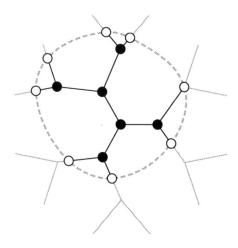

**Fig. 6.4.** The class of trees considered by Friedman [75] can be obtained by cutting connected subsets out of the geometric realization of an infinite $d$-regular tree. (• ... interior vertices, ○ ... boundary vertices)

Notice that trees are discrete analogs of *hyperbolic manifolds*. (Grids are examples of graphs which correspond to *Euclidean manifolds*.) For that reason there exists a nontrivial lower bound for the first Dirichlet eigenvalue.

**Lemma 6.10 ([75]).** *For a $d$-semiregular tree $G$ with boundary we have*

$$\lambda^\circ(G) > d - 2\sqrt{d-1} \ .$$

Friedman [75] conjectured that such trees with the Faber-Krahn property are balls centered at a vertex. Amazingly, this conjecture is false: such graphs have a more complex structure. Balls centered at a vertex do not minimize the first Dirichlet eigenvalue unless all boundary edges have length 1 [125]. Nevertheless, every tree with the Faber-Krahn-property closely resembles a ball. It looks a little bit like a "sloppily peeled onion", see Fig. 6.6. We need some definitions to describe such trees more precisely.

A *branch* $\mathrm{Br}(w, v)$ at vertex $w$ is the maximal subgraph induced by $w$, $v$ and all children $u \in V$ of $v$ (i.e. the geodesic path $(w, \ldots, u)$ contains $v$, see Fig. 6.5). The *length* $\ell(\mathrm{Br}(w, v))$ is the maximal distance $\mathrm{dist}(w, z)$, $z \in \partial V$, in $\mathrm{Br}(w, v)$. The branch is called *balanced* if $h(z)$ is the same for all boundary vertices $z \in \partial V \cap \mathrm{Br}(w, v)$, see Fig. 6.5.

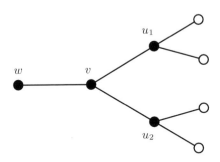

**Fig. 6.5.** A balanced branch $\mathrm{Br}(w, v)$ of length 2.7.

We say a $d$-semiregular tree with boundary $G(V^\circ \cup \partial V, E^\circ \cup \partial E)$ is *onion-shaped* if there exists a root $m \in V^\circ$ of the tree such that the following conditions are satisfied (see Fig. 6.6):

(O1) $G$ is connected.
(O2) $B(m, r) \subseteq G \subseteq B(m, r+1)$ for an $r \in \mathbb{Z}_0$ (if $|V^\circ| = 1$ then $r = 0$). Thus $|h(v_0) - h(u_0)| \le 1$ for all boundary vertices $u_0, v_0 \in \partial V$.
(O3) All boundary edges have length 1 or length $c$, where $c \in (0, 1)$ is the same for all boundary edges of length $< 1$.
(O4) If two branches $\mathrm{Br}(w_1, v_1)$ and $\mathrm{Br}(w_2, v_2)$, for $h(w_1) \ge h(w_2)$, are not balanced, then $\mathrm{Br}(w_1, v_1) \subseteq \mathrm{Br}(w_2, v_2)$.

Notice that onion-shaped trees are SLO-trees. On the other hand every a SLO-tree satisfies (O1) and (O2), but not necessarily (O3) and (O4). These hold for unweighted SLO-trees where all interior vertices have the same degree.

Using the notion of onion-shaped trees we are able to give a full characterization of trees with the Faber-Krahn property.

**Theorem 6.11 ([126]).** *A $d$-semiregular tree with boundary $G$, $d \ge 3$, has the Faber-Krahn property if and only if $G$ is onion-shaped and one of the following conditions is satisfied:*

(F0) *There is only one interior vertex, i.e. $|V^\circ| = 1$.*
(F1) *All branches of length $\ell \in (1, 2]$ are balanced (this just follows from (O3)), there is at most one balanced branch of length $\ell \in (1, 2)$, $B(m, 1) \subseteq G$, and*
   *$d \ge 5$, or*

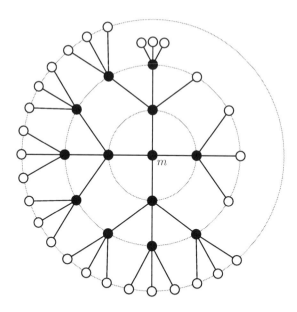

**Fig. 6.6.** Onion shaped 4-semiregular tree of volume 38.5 with Faber-Krahn property. (● ... interior vertices, ○ ... boundary vertices, $m$ ... root)

> $d = 4$ and $G \subseteq B(z, 4.5)$, or
> $d = 3$ and $G \subseteq B(z, 2.5)$.
>
> Here and in the following conditions $z$ is the midpoint of some line in $\mathcal{G}$.
> (F2) All branches of length $\ell \in (2, 3]$ are balanced, there is at most one balanced branch of length $\ell \in (2, 3)$, and
>> $d = 4$ and $B(z, 4.5) \subseteq G$, or
>> $d = 3$ and $B(z, 2.5) \subseteq G \subseteq B(z, 9.5)$.
> (F3) All branches of length $\ell \in (3, 4]$ are balanced, there is at most one balanced branch of length $\ell \in (3, 4)$, and
>> $d = 3$ and $B(z, 9.5) \subseteq G$.

$G$ is uniquely defined for a given volume up to isomorphism.

Figure 6.7 shows the 3-semiregular ball $B(z, 2.5)$, where $z$ is the midpoint of some edge. Figure 6.8 shows some regular trees of degree 3 of increasing volume with the Faber-Krahn property.

## 6.5 Rearrangements and Dirichlet Operators

The proofs of the Faber-Krahn type theorems 6.5, 6.6, 6.7, and 6.11 are lengthy and tedious. As in the continuous version of these theorems, rearrangements of the domain and the associated eigenfunction are the main tool. In the

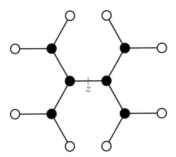

**Fig. 6.7.** The 3-semiregular ball $B(z, 2.5)$ with center $z$.
($\bullet$ ... interior vertices, $\circ$ ... boundary vertices)

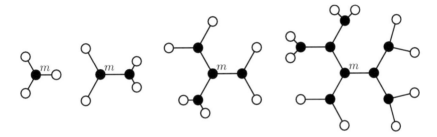

**Fig. 6.8.** Semiregular trees of degree 3 with the Faber-Krahn property.
($\bullet$ ... interior vertices, $\circ$ ... boundary vertices)

continuous case the coarea formula is used to derive the inequalities for the Rayleigh quotients (see, e.g., [29]). A powerful discrete analog of the coarea formula is not know, however, so that more elementary inequalities must be used. In the following we will therefore only present the main ideas of the proofs. The full details of the many explicit calculations can be found in [20] and [126], respectively. The basic types of rearrangement steps are *switching* and *shifting*. Related operations were used in [153].

**Lemma 6.12 (Switching).**   *Let $G(V, E)$ be a tree with boundary in some class $\mathcal{T}_\pi$. Let $(v_1, u_1), (v_2, u_2) \in E$ be edges such that $u_2$ is in the geodesic path from $v_1$ to $v_2$, but $u_1$ is not, see Fig. 6.9. Then by replacing edges $(v_1, u_1)$ and $(v_2, u_2)$ by the edges $(v_1, v_2)$ and $(u_1, u_2)$ we get a new tree $G'(V, E')$ which is also contained in $\mathcal{T}_\pi$ with the same set of boundary vertices. Moreover, we find for a function $f \in \mathcal{F}_0$*

$$\mathcal{R}_{G'}(f) \leq \mathcal{R}_G(f) \tag{6.2}$$

*whenever $f(v_1) \geq f(u_2)$ and $f(v_2) \geq f(u_1)$. Inequality (6.2) is strict if both inequalities are strict.*

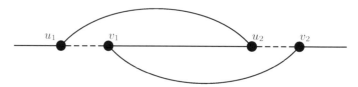

**Fig. 6.9.** Switching: edges $(v_1, u_1)$ and $(v_2, u_2)$ are replaced by edges $(v_1, v_2)$ and $(u_1, u_2)$.

*Proof.* Since by assumption $u_2$ is in the geodesic path from $v_1$ to $v_2$ and $u_1$ is not, $G'(V, E')$ again is a tree. The set of vertices does not change by construction. Moreover, since this switching does not change the degrees of the vertices, the degree sequence remains unchanged. To verify (6.2) we have to compute the effects of removing and inserting edges and get

$$\langle \mathbf{L}(G')f, f \rangle - \langle \mathbf{L}(G)f, f \rangle = \left[ (f(v_1) - f(v_2))^2 + (f(u_1) - f(u_2))^2 \right]$$
$$- \left[ (f(v_1) - f(u_1))^2 + (f(v_2) - f(u_2))^2 \right]$$
$$= 2 \left( f(u_1) - f(v_2) \right) \cdot \left( f(v_1) - f(u_2) \right)$$
$$\leq 0 ,$$

where the last inequality is strict if both inequalities $f(v_1) \geq f(u_2)$ and $f(v_2) \geq f(u_1)$ are strict. Thus the proposition follows. □

*Remark 6.13.* If $f$ is an eigenfunction of $\lambda^\circ(G)$ in Lemma 6.12 then (6.2) is strict if at least *one* of the inequalities, $f(v_1) \geq f(u_2)$ or $f(v_2) \geq f(u_1)$ is strict [20, Lemma 6].

Analogously we define *shifting*.

**Lemma 6.14 (Shifting).** *Let* $G(V, E)$ *be a tree with boundary in some graph class* $\mathcal{T}$. *Let* $(u, v_1) \in E$ *be an edge and* $v_2 \in V$ *some vertex such that* $u$ *is not in the geodesic path from* $v_1$ *to* $v_2$, *see Fig. 6.10. Then by replacing edge* $(u, v_1)$ *by the edge* $(u, v_2)$ *we get a new tree* $G'(V, E')$. *If* $v_2 \in V^\circ$ *is an interior vertex and* $d_{v_1} \geq 3$ *then the number of boundary vertices remains unchanged. Moreover, we find for a nonnegative function* $f \in \mathcal{F}_0$

$$\mathcal{R}_{G'}(f) \leq \mathcal{R}_G(f) \tag{6.3}$$

*if* $f(v_1) \geq f(v_2) \geq f(u)$. *The inequality is strict if* $f(v_1) > f(v_2)$.

Now let $n = |V|$ and $k = |V^\circ|$ denote the number of vertices and of interior vertices of $G$, respectively, and let $f$ be a nonnegative eigenfunction of the first Dirichlet eigenvalue of $G$. The degree sequence of $G$ is given by $\pi = (d_0, d_1, \ldots, d_{k-1}, d_k, \ldots, d_{n-1})$ such that the degrees $d_i$ are nondecreasing for $0 \leq i < k$, and $d_j = 1$ for $j \geq k$ (i.e., correspond to boundary vertices). We assume that the vertices of $G$, $V = \{v_0, v_1, \ldots, v_{k-1}, v_k, \ldots, v_{n-1}\}$, are numbered such that $f(v_i) \geq f(v_j)$ if $i < j$, i.e., they are sorted with respect

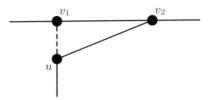

**Fig. 6.10.** Shifting: edge $(u, v_1)$ is replaced by edge $(u, v_2)$.

to $f(v)$ in nonincreasing order. We define a well-ordering $\prec$ on $V$ by $v_i \prec v_j$ if and only if $i < j$. By applying a series of switchings and shiftings

$$G = G_0 \to G_1 \to G_2 \to \ldots \to G_s = G^*$$

one can transform $G(V, E) \in \mathcal{T}_\pi$ into a graph $G^*(V, E^*) \in \mathcal{T}_\pi$ in which $\prec$ is a SLO*-ordering and where the Rayleigh quotient is nonincreasing

$$\lambda^\circ(G) = \mathcal{R}_{G_0}(f) \ge \mathcal{R}_{G_1}(f) \ge \ldots \ge \mathcal{R}_{G_s}(f) \ge \lambda^\circ(G^*) \,.$$

This procedure can be compiled as

**Algorithm Rearrange**
**Input:** Tree $G(V, E) \in \mathcal{T}_\pi$.
**Output:** Tree $G^*(V, E^*) \in \mathcal{T}_\pi$ with SLO-ordering $\prec$ and $\lambda(G^*) \le \lambda(G)$.
1: Define a well-ordering $\prec$: $v_i \prec v_j$ if and only if $i < j$.
2: Set $s \leftarrow 0$.
3: **for** $r = 0, \ldots, k - 1$ **do**
4:     **for** $i = 1, \ldots, d_r - 1$ **do**        $[\, i = 1, \ldots, d_0 \text{ if } r = 0 \,]$
5:         Set $s \leftarrow s + 1$ (increment $s$).
6:         **if** $v_s$ is not adjacent to $v_r$ **then**
7:             Select an edge $(v_r, w_r)$ such that $v_s \prec w_r$.
8:             Select an edge $(v_s, w_s)$ such that $v_s \prec w_s$ and $w_s$ is in the geodesic path from $v_r$ to $v_s$ if and only if $w_r$ is not.
9:             Apply switching such that the new graph $G_s$ has edges $(v_r, v_s)$ and $(w_r, w_s)$.
10:     **for all** $(v, v_r) \in E$ with $v_s \prec v$ **do**
11:         Apply shifting such that edge $(v, v_r)$ is replaced by edge $(v, v_{r+1})$.
12: Return $G^* = G_s$.

The difficult part for the proof of Thm. 6.7 is to show that at least one of the above inequalities is strict unless $G$ is already a SLO*-tree. For this task we need two properties of an eigenfunction $f$ of the first Dirichlet eigenvalue: every interior vertex has a child where $f$ is strictly decreasing; and subtrees at two vertices $v$ and $w$ are isomorphic if $f(v) = f(w)$. The details of this proof is rather tedious and thus omitted. The interested reader is therefore referred to the original paper [20]. The proof of Thm. 6.6 (and thus Thm. 6.5) is quite similar.

## 6.6 Perturbations and Branches

When boundary edges may have weights (as in Thm. 6.11) and hence edge lengths can be smaller than 1, we have to investigate the normal derivative of the eigenfunction $f$ of the first Dirichlet eigenvalue $\lambda^\circ(G)$ on boundary edges. For a boundary edge $vz$, $v \in V^\circ$ and $z \in \partial V$, of length $c_{vz} = 1/w_{vz}$ the normal derivative is given by $f(v)/c_{vz}$. The "average" normal derivate of all boundary edges is given by $\sum_{wz \in \partial E} f(w) / \sum_{wz \in \partial E} c_{wz}$. If we replace all boundary edges $vz$ of length $c_{vz}$ of a semiregular tree $G$ by edges of length

$$\bar{c}_{vz} = f(v) \frac{\sum_{wz \in \partial E} c_{wz}}{\sum_{wz \in \partial E} f(w)}$$

we obtain a new semiregular tree $G'$ with the same volume, $\mu_E(G') = \mu_E(G)$, where the normal derivative is the same for all these boundary edges. Moreover, a straightforward computation shows that the Rayleigh quotient is not increased, i.e., $\lambda^\circ(G) = \mathcal{R}_G(f) \geq \mathcal{R}_{G'}(f) \geq \lambda^\circ(G')$ where equality only holds if $G$ and $G'$ coincide. However, some of the boundary edges in $G'$ might become longer than 1. Then we replace all the edges $vz$ by edges $vz(\varepsilon)$ of length $c_{vz}(\varepsilon) = (1 - \varepsilon) c_{vz} + \varepsilon \bar{c}_{vz}$, where $\varepsilon \in (0, 1]$ is the same for all edges. When we make $\varepsilon$ as large as possible, i.e. (either) one edge $vz_\varepsilon$ has length $c_{vz}(\varepsilon) = 1$ or $\varepsilon = 1$, we get a new semiregular tree which we denote by $G(\varepsilon)$.

**Lemma 6.15 (Perturbation of edges, [125]).** *For the semiregular tree $G(\varepsilon)$ with boundary as constructed above we have $\mu_E(G(\varepsilon)) = \mu_E(G)$ and $\lambda^\circ(G(\varepsilon)) \leq \lambda^\circ(G_E)$. Equality holds if and only if $G(\varepsilon)$ and $G$ are isomorph.*

This lemma allows us to construct a graph with smaller first Dirichlet eigenvalue. Furthermore, we have the following immediate corollary.

**Corollary 6.16.** *If a semiregular tree has the Faber-Krahn property then the normal derivative is the same at all boundary edges of length less than 1.*

*Remark 6.17.* This corollary can alternatively be derived from analytic perturbation theory [13] for linear operators, see [75].

Notice that for all balanced branches of a $d$-semiregular tree $G$ with boundary the eigenfunction $f$ of $\lambda^\circ(G)$ is "symmetric", i.e., $f(v)$ does not change for any automorphism that acts on the vertices of such a branch. This is a consequence of the simplicity of eigenvalue $\lambda^\circ(G)$. We can use this fact to compute the eigenfunction on such a branch by means of a recursion based on (2.2).

**Lemma 6.18.** *Let $(v_0, v_1, v_2, \ldots, v_{r-1}, v_r)$ be a geodesic path in $G$ with $v_0 \in \partial V$, $v_i \in V^\circ$, and $h(v_i) = h(v_{i-1} + 1)$, for $i = 1, \ldots, r$. Let $c$ denote the length of the boundary edge $(v_0, v_1)$. If $\mathrm{Br}(v_r, v_{r-1})$ is a balanced branch then*

$$f(v_2) = ((d - 1) + (1 - \lambda^\circ(G)) c) f(v_1)/c$$
$$f(v_j) = (d - \lambda^\circ(G)) f(v_{j-1}) - (d - 1) f(v_{j-2}), \qquad j = 3, \ldots, r \, .$$

*Remark 6.19.* A symmetric eigenfunction $f$ of a ball of radius $r \in \mathbb{Z}$ additionally must satisfy $d\left(f(v_r) - f(v_{r-1})\right) = \lambda^\circ f(v_r)$. Setting $f(v_1) = c$ in Lemma 6.18 we arrive at a polynomial of degree $r - 1$ for $\lambda^\circ$. The first Dirichlet eigenvalue of such a ball is then the smallest root of this polynomial.

If we set $f(v_1) = c$, i.e., if $f$ has normal derivative 1 on all boundary edges, then we can express $f(v_j)$ by

$$f(v_j) = \alpha_j(d, \lambda^\circ(G)) + \beta_j(d, \lambda^\circ(G)) \, c \, .$$

The coefficients $\alpha_j$ and $\beta_j$ are polynomials in $d$ and $\lambda^\circ(G)$ and are determined by the recursion

$$\alpha_1 = 0, \ \alpha_2 = d - 1 \ \text{ and } \ \alpha_i = (d - \lambda^\circ)\,\alpha_{i-1} - (d - 1)\,\alpha_{j-2} \, ,$$
$$\beta_1 = 1, \ \beta_2 = 1 - \lambda^\circ \ \text{ and } \ \beta_i = (d - \lambda^\circ)\,\beta_{i-1} - (d - 1)\,\beta_{i-2} \, .$$

These coefficients have some nice properties, e.g., for fixed $d$ the roots of $\beta_k$ and $\beta_{k+1}$ as polynomials in $\lambda^\circ$ are interlaced (see [126] for details). Now it can be shown (in the tedious part of the proof in [126]) that the existence of balanced and unbalanced branches of noninteger length $\ell < k$ of a semiregular tree with the Faber-Krahn property is related to the smallest root of $\beta_k$. If this root is less than $\lambda^\circ(G)$ then no such branches can exist. However, by Lemma 6.10, $\lambda^\circ(G)$ is nontrivially bounded from below while the smallest root of $\beta_k$ tends to 0 for increasing $k$. This surprising coincidence is the reason for the conditions (F1)–(F3) in Thm. 6.11. The balls listed there have the property that their first Dirichlet eigenvalues $\lambda^\circ(B)$ are exactly the smallest roots of $\beta_2$ and $\beta_3$, respectively. The details of the proof of Thm. 6.11 are omitted and the interested reader is referred to the original paper [126].

# A

# Basic Notations

In this section we recall some notions from graph theory, linear algebra and complexity of algorithms that are used in this monograph.

## Graphs

We briefly give the terminology in graph theory needed in this monography. For standard graph theoretical terms not defined here we refer to [56, 175].

A graph $G(V, E)$ consists of a nonempty finite set $V$ called the *vertex set* and an *edge set* $E$, where an edge is an unordered pair of distinct vertices; hence we can write $x \in e$ to mean that vertex $x$ is incident with the edge $e$. For simplicity we write $xy$ (instead of $\{u, v\}$) for an edge with end-vertices $x$ and $y$.

We use $V(G)$ and $E(G)$ to denote the vertex set and edge set of $G$, respectively. We have tried to use $n = |V|$ and $m = |E|$ consistently for the respective numbers of vertices and edges ($|S|$ denotes the *cardinality* of a set $S$). Graphs as we have defined them are also referred to as *simple graphs*, since they do not have multiple edges or loops.

An edge $e = uv$ connects the vertices $u$ and $v$, and we say that $u$ and $v$ are *adjacent* or $u$ is a *neighbor* of $v$ (and vice versa). We write $v \sim u$ to express more explicitly that $v$ is adjacent to $u$ (and vice versa). In particular we use $\sum_{uv \in E}$ if we sum over all edges of a graph and $\sum_{v \sim u}$ if we sum over all vertices $v$ that are adjacent to some vertex $u$. The number of neighbors of $v$ is called the *degree* of $v$ and denoted by $d(v)$. If all the vertices of a graph $G$ have the same degree $k$, then $G$ is *k-regular*, or simply *regular*.

The *complement* $G^c$ of a graph $G$ has the same vertex set as $G$ and two vertices $u$ and $v$ are adjacent in $G^c$ if and only if they are not adjacent in $G$.

A graph is called *complete* if every pair of vertices are adjacent. We denote the complete graph with $n$ vertices by $K_n$.

A graph $H$ is a *subgraph* of $G$ if $V(H) \subseteq V(G)$ and $E(H) \subseteq E(G)$. A subgraph $H$ of $G$ is an *induced subgraph* if two vertices of $V(H)$ are adjacent if and only if they are adjacent in $G$. If $U \subseteq V(G)$, then $G[U]$ denotes the induced subgraph of $G$ with vertex set $U$. If $U$ is some set of vertices of $G$, we write $G - U$ for $G[V \setminus U]$. We write $G - v$ rather than $G - \{v\}$ and say *deletion* of vertex $v$. For a subset $F$ of $E(G)$ we write $G - F = G(V, E \setminus F)$. Instead of $G - \{e\}$ for an edge $e$ we write $G - e$ and say *deleting* edge $e$. By $G/e$ we denote the graph obtained from $G$ by *contracting* the edge $e = uv$ into a new vertex $v_e$ which becomes adjacent to all the former neighbors of both $u$ and of $v$. We delete any multiple edges or loops.

A graph $H$ that can be obtained from $G$ by a series of deletions and contractions of edges and deletions of isolated vertices is called a *minor* of $G$.

A *clique* is a subgraph that is complete. A set of vertices is *independent* if no two of its elements are adjacent.

A *path* with $k$ vertices from $u$ to $v$ in a graph is a sequence of $k$ distinct vertices starting with $u$ and ending with $v$ such that consecutive vertices are adjacent. We denote a path with $k$ vertices by $P_k$. If there is a path between any two vertices of a graph $G$, then $G$ is *connected*, otherwise *disconnected*. A maximal connected induced subgraph of $G$ is called *(connected) component* of $G$.

A *cycle* is a connected graph where every vertex has exactly two neighbors. A graph containing no cycles is called a *forest*. A connected forest is called a *tree*.

A graph $G(V, E)$ is called *k-partite* if $V$ admits a partition into $k$ classes such that vertices in the same partition class must not be adjacent. Instead of 2-partite one usually says *bipartite*. An $k$-partite graph in which every two vertices from different partition classes are adjacent is called *complete* and is denoted by $K_{n_1, \dots, n_k}$.

## Linear Algebra

We recall the main results of the linear algebra of symmetric matrices over the real numbers. For further details we refer the reader to the relevant literature, e.g. [100]. However, as most of the results are inspired by the close analogy between the continuous Laplace-Beltrami operator on Riemannian manifolds and the graph Laplacian, we will often use a different notion and terminology as we will say *function* (over a subset of $\mathbb{N}$) instead of *vector* and write $f(i)$ (or $x(i)$) instead of $x_i$ for the $i$-th component of the functions/vectors $f$ and $\mathbf{x}$, respectively.

In fact we can interpret a vector $\mathbf{x}$ with components indexed by the vertices $V$ of a given Graph $G(V, E)$ as a real-valued function $f$ on $V$, i.e. $f : V \to \mathbb{R}$. Furthermore, the set of all such functions obviously forms a vector space that is isomorphic to $\mathbb{R}^n$ and thus we can denote this vector space in abuse of language simple by $\mathbb{R}^n$. We also have the scalar product for such functions $f$ and $g$ given by $\langle f, g \rangle = \sum_{v \in V} f(x) g(x)$ and hence the space of our function forms a Hilbert space.

Let $\mathbf{A} = (A_{ij})$ be a real $n \times n$ matrix. An *eigenvalue* of $\mathbf{A}$ is a number $\lambda$ satisfying $\mathbf{A}\mathbf{x} = \lambda\mathbf{x}$ for a nonzero vector $\mathbf{x}$. Any such vector $\mathbf{x}$ is called an *eigenvector* of the matrix $\mathbf{A}$ belonging (affording) to the eigenvalue $\lambda$. Due to our convention we will say *eigenfunction* instead of eigenvector. The space of all eigenfunctions of $\mathbf{A}$ belonging to $\lambda$ together with the null function, is called the *eigenspace* $\mathcal{E}_\lambda$ of $\lambda$. The dimension of the eigenspace is called the *geometric multiplicity* of $\lambda$. The eigenvalues of $\mathbf{A}$ are the roots of the *characteristic polynomial* $\det(\mathbf{A} - \lambda\mathbf{I})$ of $\mathbf{A}$.

For a symmetric real $n \times n$ matrix $\mathbf{A}$ all eigenvalues are real and the geometric multiplicity is equal to the algebraic multiplicity of $\lambda$, i.e., the multiplicity of $\lambda$ as root of the characteristic polynomial. Furthermore there exists an orthogonal basis of the $\mathbb{R}^n$ consisting of eigenfunctions of $\mathbf{A}$.

The *spectrum* of a matrix is the list of its eigenvalues together with their multiplicities. The *spectral radius* $\rho(\mathbf{A})$ of a matrix is the maximum of the absolute values of its eigenvalues.

The *trace* of a square matrix $\mathbf{A}$ is the sum of the diagonal entries and is denoted by $\mathrm{tr}(\mathbf{A})$. The trace of a square matrix is also equal to the sum of its eigenvalues, i.e., $\mathrm{tr}(\mathbf{A}) = \sum_{i=1}^{n} a_{ii} = \sum_{i=1}^{n} \lambda_i$.

A matrix $\mathbf{B}$ is called a *principal submatrix* of a symmetric matrix $\mathbf{A}$ if it is obtained by removing corresponding rows and columns from $\mathbf{A}$.

## Algorithms and Their Complexity

In the analysis of an algorithm first of all we are interested in its *complexity*, which is measured by the number of elementary operations that it requires. The complexity of an algorithm depends on the size of its input. An algorithm is said to be an $\mathcal{O}(g(N))$ *algorithm* of its input size $N$ for some function $g(\cdot)$ if the running time never exceeds $c\,g(N)$ for some positive constant $c$. An algorithm is a *polynomial algorithm* if $g(N)$ is a polynomial in $N$.

There are many interesting algorithmic problems concerning graphs for which no polynomial algorithm are known. Many of those problems belong to the class of *NP-complete* problems. For a detailed introduction to the class of NP-complete problems, see [80].

A problem is a *decision problem* if it requires the answer "yes" or "no". A problem is understood as a family of *instances*. For example, we consider the *Hamilton cycle problem*: given a graph, decide whether or not it has a Hamilton cycle. Every graph provides an instance of this problem.

A decision problem $S$ belongs to the complexity class $P$ if and only if there exists a polynomial algorithm which, given any instance of $S$, produces answer "yes" or "no" such that the answer of the algorithm on input $x$ is "yes" if and only if $x$ is a "yes" instance for $S$.

A decision problem belongs to the complexity class $NP$ if, for every "yes" instance of the problem, there exists a short "proof", called a *certificate*, of polynomial size such that, using the certificate, one can verify in polynomial time that the instance is indeed a "yes" instance.

Given a pair of decision problems $S$ and $T$, we say that $S$ is *polynomially reducible* to $T$ if there is a polynomial algorithm $\mathcal{A}$ that transforms an instance $x$ of $S$ into an instance $\mathcal{A}(x)$ of $T$ such that the second instance has the same answer as the first one. That is, $x$ is a "yes" instance of $S$ if and only if $\mathcal{A}(x)$ is a "yes" instance of $T$.

A decision problem is *NP-hard* if all problems in NP can be polynomially reduced this problem. If the problem is NP-hard and also belongs to NP then it is *NP-complete*. Polynomial transformations are transitive. Hence, in order to prove that a problem $W$ is NP-hard, it is sufficient to prove that there is some NP-complete problem which is polynomial reducible to $W$.

# B

# Eigenfunctions Used in Figures

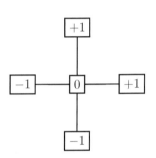

**Fig. 3.1.** Laplacian **L**, $\lambda_2 = 1$ (multiplicity $r = 3$).

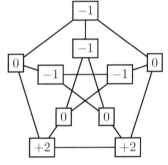

**Fig. 3.2.** Laplacian **L**, $\lambda_2 = 2$ (multiplicity $r = 5$).

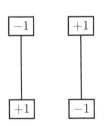

**Fig. 3.3.** Laplacian **L**, $\lambda_3 = 2$ (multiplicity $r = 2$).

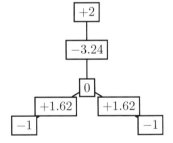

**Fig. 3.4.** Laplacian **L**, $\lambda_5 = (3 + \sqrt{5})/2$ (multiplicity $r = 2$; numbers rounded to 3 digits).

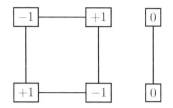

**Fig. 3.5.** Laplacian **L**, $\lambda_6 = 4$ (simple).

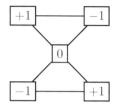

**Fig. 3.7.** Laplacian **L**, $\lambda_3 = 2$ (multiplicity $r = 2$).

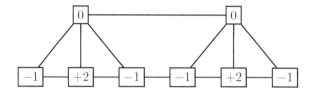

**Fig. 3.6.** Laplacian **L**, $\lambda_5 = 4$ (multiplicity $r = 2$).

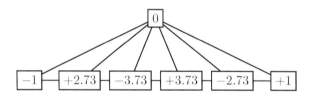

**Fig. 3.8.** Laplacian **L**, $\lambda_6 = 3 + \sqrt{3}$ (simple; numbers rounded to 3 digits).

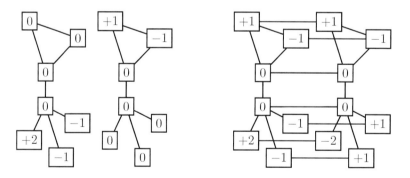

**Fig. 4.3.** Laplacian **L**, $\lambda_3(G) = 1$ (multiplicity $r = 2$), $\lambda_5(G) = 3$ (simple), and $\lambda_7(G \square K_2) = 3$ (multiplicity $r = 3$).

# C

# List of Symbols

| | |
|---|---|
| $\mathbb{R}$ | set of real numbers |
| $\mathbb{R}^n$ | real vector space of dimension $n$ |
| $\mathbb{S}^{n-1}$ | unit sphere in $\mathbb{R}^n$ |
| $|S|$ | cardinality of set $S$ |
| $S^\perp$ | orthogonal complement of set $S \subset \mathbb{R}$ |
| $\langle \mathbf{x}, \mathbf{y} \rangle$ | scalar product of vectors $\mathbf{x}$ and $\mathbf{y}$ |
| $G(V, E)$ | graph with vertex set $V$ and edge set $E$ |
| $V(G)$ | vertex set of $G$ |
| $n$ | number of vertices $(n = |V|)$ |
| $V^\circ, \partial V$ | set of interior and boundary vertices of graph $G$ with boundary |
| $E(G)$ | edge set of graph $G$ |
| $m$ | number of vertices $(m = |E|)$ |
| $E^\circ, \partial E$ | set of interior and boundary edges of graph $G$ with boundary |
| $d(v)$ | degree of vertex $v$ |
| $uv$ | edge with incident vertices $u$ and $v$ |
| $y \sim x$ | $y$ is adjacent to $x$ (and vice versa) |
| $\mathcal{G}$ | geometric realization of graph $G$ |
| $\mathrm{dist}(u, v)$ | geodesic distance between $u, v \in \mathcal{G}$ |
| $h(v)$ | height of a vertex $v$ in rooted tree |
| $\mathrm{Br}(w, v)$ | branch of a tree spanned by edge $wv$ |
| $B(x, r)$ | ball in $\mathcal{G}$ with center $x$ a radius $r$ |
| $\mathrm{Aut}(G)$ | automorphism group of graph $G$ |
| $G[U]$ | induced subgraph of $G$ with vertex set $U$ |
| $G^c$ | complement of graph $G$ |
| $G - e$ | deletion of edge $e$ |
| $G - v$ | deletion of vertex $v$ |
| $G/e$ | contraction of edge $e$ |
| $G\{W\}$ | reduced graph |
| $G \square H$ | Cartesian product of two graphs $G$ and $H$ |
| $G + H$ | disjoint union of two graphs $G$ and $H$ |
| $G * H$ | join of two graphs $G$ and $H$ |

| | |
|---|---|
| $K_n$ | complete graph with $n$ vertices |
| $K_{n_1,\ldots,n_k}$ | complete $k$-partite graph |
| $K_2^d$ | $d$-dimensional hypercube |
| $P_k$ | path with $k$ vertices |
| $\nabla$ | incidence matrix of a graph of dimension $|E| \times |V|$ |
| $\mathbf{A}(G)$ | adjacency matrix of graph $G$ |
| $\mathbf{D}(G)$ | degree matrix of graph $G$ |
| $\mathbf{L}(G)$ | Laplacian of graph $G$ |
| $\mathbf{L}_w(G)$ | Laplacian of a weighted graph $G$ with weights $w$ |
| $\mathbf{L}^\circ(G)$ | Dirichlet matrix of graph $G$ with boundary |
| $\mathbf{M}(G)$ | generalized Laplacian of graph $G$ |
| $\mathbf{A}, \mathbf{L}, \mathbf{M}$ | respective shorthand for adjacency matrix, graph Laplacian, and generalized graph Laplacian, when graph is clear from context |
| $\lambda_i$ | $i$-th eigenvalue of (generalized) graph Laplacian $\mathbf{L}$ (or $\mathbf{M}$) |
| $\mathcal{E}_\lambda$ | eigenspace of eigenvalue $\lambda$ |
| $\lambda^\circ(G)$ | lowest Dirichlet eigenvalue of graph $G$ |
| $\mathfrak{S}(f)$ | number of strong nodal domains of function $f$ |
| $\mathfrak{W}(f)$ | number of weak nodal domains of function $f$ |
| $\mathbf{A} \otimes \mathbf{B}$ | Kronecker product of two matrices $\mathbf{A}$ and $\mathbf{B}$ |
| $\mathbf{I}$ | identity matrix |
| $\mathbf{J}$ | matrix of all ones |
| $\rho(\mathbf{A})$ | spectral radius of matrix $\mathbf{A}$ |
| $\varphi_I(v)$ | Walsh function |
| $\mathcal{O}(\cdot)$ | Landau symbol |
| $\mathcal{F}_0$ | set of all functions $f$ on vertex set $V = V^\circ \cup \partial V$ with $f|_{\partial V} = 0$ |

# References

[1] C. J. Alpert, A. B. Kahng, and S. Yao. Spectral partitioning with multiple eigenvectors. *Discrete Appl. Math.*, 90:3–26, 1999.

[2] E. Angel and V. Zissimopoulos. On the classification of NP-complete problems in terms of their correlation coefficient. *Discr. Appl. Math.*, 99:261–277, 2000.

[3] V. I. Arnol'd, M. I. Vishik, Y. S. Il'yashenko, A. S. Kalashnikov, V. A. Kondrat'ev, S. N. Kruzhkov, E. M. Landis, V. M. Millionshchikov, O. A. Olejnik, A. F. Filippov, and M. A. Shubin. Some unsolved problems in the theory of differential equations and mathematical physics. *Russian Math. Surveys*, 44(4):157–171, 1989.

[4] B. Aspvall and J. R. Gilbert. Graph coloring using eigenvalue decomposition. *SIAM J. Alg. Disc. Meth.*, 5(4):526–538, 1984.

[5] F. M. Atay, T. Bıyıkoğlu, and J. Jost. Synchronization of networks with prescribed degree distribution. *IEEE Trans. Circuits Syst. I: Fundamental Theory and Applications*, 53(1):92–98, 2006.

[6] A. T. Balaban, editor. *Chemical Applications of Graph Theory*. Academic Press, London, 1984.

[7] R. Bapat, S. Kirkland, and S. Pati. The perturbed Laplacian matrix of a graph. *Linear Multilinear Algebra*, 49(3):219–242, 2001.

[8] R. B. Bapat and S. Pati. Algebraic connectivity and the characteristic set of a graph. *Linear Multilinear Algebra*, 45(2–3):247–273, 1998.

[9] S. Barik and S. Pati. On algebraic connectivity and spectral integral variations of graphs. *Linear Algebra Appl.*, 397:209–222, 2005.

[10] F. Barioli, S. Fallat, and L. Hogben. A variant on the graph parameters of Colin de Verdière: Implications to the minimum rank of graphs. *Elec. J. Lin. Alg.*, 13:387–404, 2005.

[11] J. Barnes, S. Dokov, R. Acevedoa, and A. Solomon. A note on distance matrices yielding elementary landscapes for the TSP. *J. Math. Chem.*, 31:233–235, 2002.

[12] J. W. Barnes, B. Dimova, S. P. Dokov, and A. Solomon. The theory of elementary landscapes. *Appl. Math. Lett.*, 16:337–343, 2003.

[13] H. Baumgärtel. *Analytic Perturbation Theory for Matrices and Operators.* Birkhäuser, Basel, 1985.

[14] L. W. Beineke, R. J. Wilson, and P. J. Cameron, editors. *Topics in Algebraic Graph Theory*, volume 102 of *Encyclopedia of Mathematics and Its Applications.* Cambridge University Press, 2004.

[15] M. Belkin and P. Niyogi. Laplacian eigenmaps and spectral techniques for embedding and clustering. In *Advances in Neural Information Processing Systems 14 (NIPS 2001)*, pages 585–591, Cambridge, 2002. MIT Press.

[16] G. Benkö, C. Flamm, and P. F. Stadler. A graph-based toy model of chemistry. *J. Chem. Inf. Comput. Sci.*, 43:1085–1093, 2003.

[17] N. Biggs. *Algebraic Graph Theory.* Cambridge University Press, Cambridge UK, 2nd edition, 1994.

[18] K. Binder and A. P. Young. Spin glasses: Experimental facts, theoretical concepts, and open questions. *Rev. Mod. Phys.*, 58:801–976, 1986.

[19] T. Bıyıkoğlu. A discrete nodal domain theorem for trees. *Lin. Algebra Appl.*, 360:197–205, 2003.

[20] T. Bıyıkoğlu and J. Leydold. Faber-Krahn type inequalities for trees. *J. Comb. Theory B*, 2006. to appear.

[21] T. Bıyıkoğlu, W. Hordijk, J. Leydold, T. Pisanski, and P. F. Stadler. Graph Laplacians, nodal domains, and hyperplane arrangements. *Lin. Algebra Appl.*, 390:155–174, 2004.

[22] T. Bıyıkoğlu, J. Leydold, and P. F. Stadler. Nodal domain theorems and bipartite subgraphs. *Elec. J. Lin. Alg.*, 13:344–351, 2005.

[23] A. Björner, L. Lovász, and P. W. Shor. Chip-firing games on graphs. *Eur. J. Comb.*, 12(4):283–291, 1991.

[24] D. Bonchev and D. H. Rouvray, editors. *Chemical Graph Theory: Introduction and Fundamentals.* Abacus/Gordon and Breach, London, 1991.

[25] A. J. Bray and M. A. Moore. Metastable states in spin glasses with short-ranged interactions. *J. Phys. C*, 14:1313–1327, 1981.

[26] R. A. Brualdi and B. L. Shader. *Matrices of Sign-solvable Linear Systems*, volume 116 of *Cambridge Tracts in Mathematics.* Cambridge Univ. Press, Cambridge, UK, 1995.

[27] R. Bürger. *The Mathematical Theory of Selection, Recombination, and Mutation.* Wiley, Chichester, UK, 2000.

[28] A. Chan and C. D. Godsil. Symmetry and eigenvectors. In G. Hahn and G. Sabidussi, editors, *Graph Symmetry*, volume 497 of *NATO ASI Series, Serie C: Mathematical and Physical Sciences*, pages 75–106. Kluwer Academic Publishers, 1997.

[29] I. Chavel. *Eigenvalues in Riemannian Geometry.* Academic Press, Orlando Fl., 1984.

[30] W.-K. Chen. On vector spaces associated with a graph. *SIAM J. Appl. Math.*, 20:525–529, 1971.

[31] S.-Y. Cheng. Eigenfunctions and nodal sets. *Comment. Math. Helvetici*, 51:43–55, 1976.

[32] F. Chung and R. B. Ellis. A chip-firing game and Dirichlet eigenvalues. *Discrete Math.*, 257(2-3):341–355, 2002. Kleitman and combinatorics: a celebration (Cambridge, MA, 1999).

[33] F. Chung, L. Lu, and V. Vu. Eigenvalues of random power law graphs. *Ann. Comb.*, 7:21–33, 2003.

[34] F. Chung, L. Lu, and V. Vu. Spectra of random graphs with given expected degrees. *Proc. Natl. Acad. Sci. USA*, 100:6313–6318, 2003.

[35] F. R. K. Chung. *Spectral Graph Theory*, volume 92 of *CBMS*. American Mathematical Society, 1997.

[36] V. Chvátal and P. L. Hammer. Set–packing and threshold graphs. Res. report, Comp. Sci. Dept. Univ. of Waterloo, Ontario CORR 73–21, 1973.

[37] B. Codenotti and L. Margara. Local properties of some NP-complete problems. Technical Report TR 92-021, International Computer Science Institute, Berkeley, CA, 1992.

[38] Y. Colin de Verdière. Sur un nouvel invariant des graphes et un critère de planarité. *J. Comb. Theory, Ser. B*, 50:11–21, 1990.

[39] Y. Colin de Verdière. Multiplicités des valeurs propres Laplaciens discrets et laplaciens continus. *Rendiconti di Matematica*, 13:433–460, 1993. (French).

[40] Y. Colin de Verdière. On a new graph invariant and a criterion for planarity. In *Graph Structure Theory*, volume 147 of *Contemporary Mathematics*, pages 137–147. Amercian Mathematical Society, 1993.

[41] Y. Colin de Verdière. *Spectres de Graphes*. Number 4 in Cours Spécialisés. Société Mathématique de France, 1998.

[42] D. G. Corneil, Y. Perl, and L. K. Stewart. Cographs: recognition, application and algorithms. *Congressus Numerantium*, 43:249–258, 1984.

[43] D. G. Corneil, Y. Perl, and L. K. Stewart. A linear recognition algorithm for cographs. *SIAM J. Comput.*, 14:926–934, 1985.

[44] R. Courant and D. Hilbert. *Methods of Mathematical Physics, Vol. 1*. Interscience, New York, 1953.

[45] D. M. Cvetković, M. Doob, and H. Sachs. *Spectra of Graphs – Theory and Applications*, volume New York. Academic Press, 1980.

[46] D. M. Cvetković, M. Doob, I. Gutman, and A. Torgašev. *Recent Results in the Theory of Graph Spectra*, volume 36 of *Annals of Discrete Mathematics*. North Holland, Amsterdam, New York, Oxford, Tokyo, 1988.

[47] D. M. Cvetković, P. Rowlinson, and S. Simić. *Eigenspaces of Graphs*, volume 66 of *Encyclopdia of Mathematics and its Applications*. Cambrigdge University Press, Cambridge, UK, 1997.

[48] E. B. Davies, J. Leydold, and P. F. Stadler. Discrete nodal domain theorems, 2000. arxiv:math.SP/0009120.

[49] E. B. Davies, G. M. L. Gladwell, J. Leydold, and P. F. Stadler. Discrete nodal domain theorems. *Lin. Algebra Appl.*, 336:51–60, 2001.

[50] V. M. de Oliveira, J. F. Fontanari, and P. F. Stadler. Metastable states in high order short-range spin glasses. *J. Phys. A: Math. Gen.*, 32: 8793–8802, 1999.

[51] B. Derrida. Random energy model: Limit of a family of disordered models. *Phys. Rev. Lett.*, 45:79–82, 1980.

[52] B. Derrida. Random-energy model: An exactly solvable model of disorderes systems. *Phys. Rev. B*, 24:2613–2626, 1981.

[53] B. Derrida and E. Gardner. Metastable states of a spin glass chain at 0 temperature. *J. Physique*, 47:959–965, 1986.

[54] L. Devroye. *Non-Uniform Random Variate Generation.* Springer-Verlag, New-York, 1986.

[55] P. Diaconis. *Group Representations in Probability and Statistics.* Inst. of Math. Stat., Hayward, CA, 1989.

[56] R. Diestel. *Graph Theory*, volume 173 of *Graduate Texts in Mathematics.* Springer-Verlag, Berlin, 2nd edition, 2000.

[57] C. Ding and H. Zha. *Spectral Clustering, Ordering and Ranking: Statistical Learning with Matrix Factorizations.* Springer, New York, 2007. Available July 2007, ISBN: 0-387-30448-7.

[58] A. W. M. Dress and D. S. Rumschitzky. Evolution on sequence space and tensor products of representation spaces. *Acta Appl. Math.*, 11: 103–115, 1988.

[59] A. M. Duval and V. Reiner. Perron-Frobenius type results and discrete versions of nodal domain theorems. *Lin. Algebra Appl.*, 294:259–268, 1999.

[60] H. Edelsbrunner. *Algorithms in Combinatorial Geometry*, volume 10 of *EATCS Monographs on Theoretical Computer Science.* Springer-Verlag, 1987.

[61] G. Faber. Beweis, daß unter allen homogenen Membranen von gleicher Fläche und gleicher Spannung die kreisförmige den tiefsten Grundton gibt. *Sitz. bayer. Akad. Wiss.*, pages 169–172, 1923.

[62] I. Faria. Permanental roots and the star degree of a graph. *Lin. Algebra Appl.*, 64:255–265, 1985.

[63] F. F. Ferreira, J. F. Fontanari, and P. F. Stadler. Landscape statistics of the low autocorrelated binary string problem. *J. Phys. A: Math. Gen.*, 33:8635–8647, 2000.

[64] M. Fiedler. Bounds for eigenvalues of doubly stochastic matrices. *Linear Algebra Appl.*, 5:299–310, 1972.

[65] M. Fiedler. Algebraic connectivity of graphs. *Czechoslovak Math. J.*, 23:298–305, 1973.

[66] M. Fiedler. Eigenvectors of acyclic matrices. *Czechoslovak Math. J.*, 25: 607–618, 1975.

[67] M. Fiedler. A property of eigenvectors of nonnegative symmetric matrices and its application to graph theory. *Czechoslovak Math. J.*, 25: 619–633, 1975.

[68] M. Fiedler. Laplacian of graphs and algebraic connectivity. In *Combinatorics and graph theory*, volume 25 of *Banach Cent. Publ.*, pages 57–70. Banach Cent. Publ., Warsaw, 1989.

[69] N. J. Fine. The generalized Walsh functions. *Trans. Am. Math. Soc.*, 69:66–77, 1950.

[70] M. E. Fisher. On hearing the shape of a drum. *J. Comb. Theory*, 1: 105–125, 1966.

[71] G. S. Fishman. *Monte Carlo. Concepts, Algorithms, and Applications.* Springer Series in Operations Research. Springer, New York, 1996.

[72] W. Fontana, P. F. Stadler, E. G. Bornberg-Bauer, T. Griesmacher, I. L. Hofacker, M. Tacker, P. Tarazona, E. D. Weinberger, and P. Schuster. RNA folding and combinatory landscapes. *Phys. Rev. E*, 47:2083–2099, 1993.

[73] J. J. Forman, P. A. Clemons, S. L. Schreiber, and S. J. Haggarty. SpectralNet — an application for spectral graph analysis and visualization. *BMC Bioinformatics*, 6:260, 2006. URL http://www.biomedcentral.com/1471-2105/6/260.

[74] P. Fowler, T. Pisanski, and J. Shawe-Taylor. Molecular graph eigenvectors for molecular coordinates. In R. Tamassia and I. Tollis, editors, *Graph Drawing*, volume 894 of *Lecture Notes in Computer Science*, pages 282–285. Springer-Verlag, Berlin, 1995.

[75] J. Friedman. Some geometric aspects of graphs and their eigenfunctions. *Duke Math. J.*, 69(3):487–525, March 1993.

[76] J. Friedman and J.-P. Tillich. Calculus on graphs, 2004. arxiv:cs.DM/0408028.

[77] J. Friedman and J.-P. Tillich. Wave equations for graphs and the edge-based Laplacian. *Pacific J. Math.*, 216(2):229–266, 2004.

[78] F. R. Gantmacher and M. G. Krein. *Oscillation matrices and kernels and small vibrations of mechanical systems.* AMS Chelsea Publishing, Providence, RI, revised edition, 2002. Translation based on the 1941 Russian original.

[79] R. García-Pelayo and P. F. Stadler. Correlation length, isotropy, and meta-stable states. *Physica D*, 107:240–254, 1997.

[80] M. R. Garey and D. S. Johnson. *Computers and Intractability. A Guide to the Theory of NP-Completeness.* W. H. Freeman and Company, San Francisco, 1979.

[81] G. M. L. Gladwell and H. Zhu. Courant's nodal line theorem and its discrete counterparts. *Quart. J. Mech. Appl. Math.*, 55(1):1–15, 2002.

[82] G. M. L. Gladwell and H. Zhu. The Courant-Herrmann conjecture. *ZAMM, Z. Angew. Math. Mech.*, 83(4):275–281, 2003.

[83] C. D. Godsil. *Algebraic Combinatorics.* Chapman & Hall, New York, 1993.

[84] C. D. Godsil. Eigenpolytopes of distance-regular graphs. *Can. J. Math.*, 50(4):739–755, 1998.

[85] C. D. Godsil and G. F. Royle. *Algebraic Graph Theory.* Springer, Heidelberg, 2001.

[86] C. Gordon, D. L. Webb, and S. Wolpert. One cannot hear the shape of a drum. *Bull. Am. Math. Soc., New Ser.*, 27:134–138, 1992.

[87] A. Graovac, D. Plavšić, M. Kaufman, T. Pisanski, and E. C. Kirby. Application of the adjacency matrix eigenvectors method to geometry determination of toroidal carbon molecules. *J. Chem. Phys.*, 113:1925–1931, 2000.

[88] R. Grone, R. Merris, and V. Sunder. The Laplacian spectrum of a graph. *SIAM J. Matrix Anal. Appl.*, 11:218–238, 1990.

[89] D. J. Gross and M. Mèzard. The simplest spin glass. *Nucl. Phys. B*, 240:431–452, 1984.

[90] L. K. Grover. Local search and the local structure of NP-complete problems. *Oper. Res. Lett.*, 12:235–243, 1992.

[91] S. Guattery and G. L. Miller. On the quality of spectral separators. *SIAM J. Matrix Anal. Appl.*, 19:701–719, 1998.

[92] W. Hackbusch. *Elliptic Differential Equations: Theory and Numerical Treatment*, volume 18 of *Springer Series in Computational Mathematics*. Springer-Verlag, Berlin, 1992. translated from German by R. Fadiman and P. D. F. Ion.

[93] P. L. Hammer and A. K. Kelmans. Laplacian spectra and spanning trees of threshold graphs. *Discrete Appl. Math.*, 65:255–273, 1996.

[94] F. Harary. *Graph theory.* Addison-Wesley, Reading MA, 1969.

[95] D. Heidrich, W. Kliesch, and W. Quapp. *Properties of Chemically Interesting Potential Energy Surfaces*, volume 56 of *Lecture Notes in Chemistry.* Springer-Verlag, Berlin, 1991.

[96] J. Hofbauer and K. Sigmund. *Evolutionary Games and Population Dynamics.* Cambridge University Press, Cambridge U.K., 1998.

[97] L. Hogben. Spectral graph theory and the inverse eigenvalue problem of a graph. *Electron. J. Linear Algebra*, 14:12–31, 2005.

[98] W. Hordijk and P. F. Stadler. Amplitude spectra of fitness landscapes. *Adv. Complex Syst.*, 1:39–66, 1998.

[99] W. Hörmann, J. Leydold, and G. Derflinger. *Automatic Nonuniform Random Variate Generation.* Springer-Verlag, Berlin Heidelberg, 2004.

[100] R. A. Horn and C. R. Johnson. *Matrix Analysis. Reprinted with corrections.* Cambridge University Press, 1990.

[101] E. Hückel. Quantentheoretische Beiträge zum Benzolproblem. *Z. Phys.*, 70:204–286, 1931.

[102] W. Imrich and S. Klavžar. *Product Graphs: Structure and Recognition.* Wiley, New York, 2000.

[103] M. Ipsen and A. S. Mikhailov. Evolutionary reconstruction of networks. *Phys. Rev. E*, 66(4):046109, 2, 2002.

[104] F. John. *Partial Differential Equations*, volume 1 of *Applied Mathematical Science.* Springer, New York, 1971.

[105] J. Jost and M. P. Joy. Spectral properties and synchronization in coupled map lattices. *Phys. Rev. E (3)*, 65:016201, 9, 2002.

[106] M. Kac. Can one hear the shape of a drum? *Am. Math. Mon.*, 73(4, Part II):1–23, 1966.

[107] В. Н. Карпушкин. Точные по порядку оцекни числа компонент дополнения к нулям гармонических полиномов. Функцион. аналис и его прил., 19:55–60, 1985. (Russian).

[108] В. Н. Карпушкин. О топологий нулей совственних функций. Функцион. аналис и его прил., 23:59–60, 1989. (Russian).

[109] A. Katsuda and H. Urakawa. The first eigenvalue of the discrete Dirichlet problem for a graph. *J. Comb. Math. Comb. Comput.*, 27:217–225, 1998.

[110] A. Katsuda and H. Urakawa. The Faber-Krahn type isoperimetric inequalities for a graph. *Tohoku Math. J., II. Ser.*, 51(2):267–281, 1999.

[111] S. Kirkland. Constructions for type I trees with nonisomorphic Perron branches. *Czech. Math. J.*, 49(3):617–632, 1999.

[112] S. Kirkland. An upper bound on algebraic connectivity of graphs with many cutpoints. *Electron. J. Linear Algebra*, 8:94–109, 2001.

[113] S. Kirkland and S. Fallat. Perron components and algebraic connectivity for weighted graphs. *Linear Multilinear Algebra*, 44(2):131–148, 1998.

[114] S. Kirkland, M. Neumann, and B. L. Shader. Characteristic vertices of weighted trees via Perron values. *Linear Multilinear Algebra*, 40(4): 311–325, 1996.

[115] S. J. Kirkland and M. Neumann. On algebraic connectivity as a function of an edge weight. *Linear Multilinear Algebra*, 52(1):17–23, 2004.

[116] S. J. Kirkland, M. Neumann, and B. L. Shader. Bounds on the subdominant eigenvalue involving group inverses with applications to graphs. *Czech. Math. J.*, 48(1):1–20, 1998.

[117] Y. Koren. Drawing graphs by eigenvectors: theory and practice. *Comp. & Math. Appl.*, 49:1867–1888, 2005.

[118] E. Krahn. Über eine von Rayleigh formulierte Minimaleigenschaft des Kreises. *Math. Ann.*, 94:97–100, 1925.

[119] M. Krivelevich and B. Sudakov. The largest eigenvalue of sparse random graphs. *Combin. Probab. Comput.*, 12(1):61–72, 2003.

[120] J. R. Kuttler and V. G. Sigillito. Eigenvalues of the Laplacian in two dimensions. *SIAM Review*, 26(2):163–193, 1984.

[121] I. Lászlo and A. Rassat. The geometric structure of deformed nanotubes and the topological coordinates. *J. Chem. Inf. Comput. Sci.*, 43:519–524, 2003.

[122] I. Lászlo, A. Rassat, P. W. Fowler, and A. Graovac. Topological coordinates for toroidal structures. *Chem. Phys. Lett.*, 342:369–374, 2001.

[123] H. Lewy. On the minimum number of domains in which the nodal lines of spherical harmonics divide the sphere. *Comm. partial Diff. Eqns.*, 2: 1233–1244, 1977.

[124] J. Leydold. On the number of nodal domains of spherical harmonics. *Topology*, 35:301–321, 1996.

[125] J. Leydold. A Faber-Krahn-type inequality for regular trees. *GAFA, Geom. Funct. Anal.*, 7(2):364–378, 1997.

[126] J. Leydold. The geometry of regular trees with the Faber-Krahn property. *Discrete Math.*, 245(1–3):155–172, 2002.

[127] Y. Li and L. Guan. Some remarks for discrete versions of nodal domain theorems. *J. Math. Res. Exposition*, 23(2):275–278, 2003.

[128] L. Lovász. Spectra of graphs with transitive groups. *Periodica Math. Hung.*, 6:191–195, 1975.

[129] L. Lovász and A. Schrijver. A Borsuk theorem for antipodal links and a spectral characterization of linklessly embeddable graphs. *Proc. Amer. Math. Soc.*, 126:1275–1285, 1998.

[130] L. Lovász and A. Schrijver. On the null space of a Colin de Verdière matrix. *Ann. Inst. Fourier*, 49(3):1017–1026, 1999.

[131] S. Maslov. Measures of globalization based on cross-correlations of world financial indices. *Physica A*, 301:397–406, 2001.

[132] R. Merris. Characteristic vertices of trees. *Linear Multilinear Algebra*, 22(2):115–131, 1987.

[133] R. Merris. Laplacian matrices of graphs: A survey. *Lin. Algebra Appl.*, 197–198:143–176, 1994.

[134] R. Merris. Laplacian graph eigenvectors. *Lin. Algebra Appl.*, 278:221–236, 1998.

[135] M. Mézard, G. Parisi, and M. Virasoro. *Spin Glass Theory and Beyond*. World Scientific, Singapore, 1987.

[136] P. G. Mezey. *Potential Energy Hypersurfaces*. Elsevier, Amsterdam, 1987.

[137] B. Mohar. The Laplacian spectrum of graphs. In Y. Alavi, G. Chartrand, O. Ollermann, and A. Schwenk, editors, *Graph Theory, Combinatorics, and Applications*, pages 871–898, New York, 1991. John Wiley and Sons, Inc.

[138] B. Mohar. Some applications of Laplace eigenvalues of graphs. In G. Hahn and G. Sabidussi, editors, *Graph Symmetry: Algebraic Methods and Applications*, volume 497 of *NATO ASI Series C: Mathematical and Physical Sciences*, pages 225–275. Kluwer Academic Publishers, 1997.

[139] B. Mohar and S. Poljak. Eigenvalues in combinatorial optimization. In R. A. Brualdi, S. Friedland, and V. Klee, editors, *Combinatorial and Graph-Theoretical Problems in Linear Algebra*, volume 50 of *IMA Volumes in Mathematics and Its Applications*, pages 107–151, Berlin, 1993. Springer-Verlag.

[140] J. J. Molitierno. The spectral radius of submatrices of Laplacian matrices for trees and its comparison to the Fiedler vector. *Linear Algebra Appl.*, 406:253–271, 2005.

[141] H. Müller. private communication, 2001.

[142] R. Palmer. Optimization on rugged landscapes. In A. S. Perelson and S. A. Kauffman, editors, *Molecular Evolution on Rugged Landscapes: Proteins, RNA, and the Immune System*, pages 3–25. Addison Wesley, Redwood City, CA, 1991.

[143] S. Pati. The third smallest eigenvalue of the Laplacian matrix. *Electron. J. Linear Algebra*, 8:128–139, 2001.

[144] T. Pisanski and J. Shawe-Taylor. Characterising graph drawing with eigenvectors. *J. Chem. Inf. Comput. Sci.*, 40:567–571, 2000.

[145] T. Pisanski and A. Žitnik. Representations of graphs and maps. In J. Gross and T. Tucker, editors, *Topological Graph Theory*, Encyclopedia of Mathematics. Cambridge University Press, 2005. in preparation.

[146] A. Pothen, H. D. Simon, and K.-P. Liou. Partitioning sparse matrices with eigenvectors of graphs. *SIAM J. Matrix Anal. Appl.*, 11:430–452, 1990.

[147] D. L. Powers. Graph partitioning by eigenvectors. *Lin. Algebra Appl.*, 101:121–133, 1988.

[148] A. R. Pruss. Discrete convolution-rearrangement inequalities and the Faber-Krahn inequality on regular trees. *Duke Math. J.*, 91(3):463–514, 1998.

[149] C. M. Reidys and P. F. Stadler. Combinatorial landscapes. *SIAM Review*, 44:3–54, 2002.

[150] H. Rieger. The number of solutions of the Thouless-Anderson-Palmer equations for $p$-spin interaction spin glasses. *Phys. Rev. B*, 46:14655–14661, 1992.

[151] D. Rockmore, P. Kostelec, W. Hordijk, and P. F. Stadler. Fast Fourier transform for fitness landscapes. *Appl. Comput. Harmonic Anal.*, 12: 57–76, 2002.

[152] R. Roth. On the eigenvectors belonging to the minimum eigenvalue of an essentially nonnegative symmetric matrix with bipartite graph. *Lin. Algebra Appl.*, 118:1–10, 1989.

[153] P. Rowlinson. Graph perturbations. In *Surveys in Combinatorics, Proc. 13th Br. Comb. Conf. Guildford/UK 1991*. Lond. Math. Soc., 1991.

[154] J. Shi and J. Malik. Normalized cuts and image segmentation. *IEEE Trans. Pattern Analysis Machine Intel.*, 22:888–905, 2000.

[155] A. Solomon, J. W. Barnes, S. P. Dokov, and R. Acevedo. Weakly symmetric graphs, elementary landscapes, and the TSP. *Appl. Math. Lett.*, 16:401–407, 2003.

[156] B. M. R. Stadler and P. F. Stadler. Generalized topological spaces in evolutionary theory and combinatorial chemistry. *J. Chem. Inf. Comput. Sci.*, 42:577–585, 2002. Proceedings MCC 2001, Dubrovnik.

[157] P. F. Stadler. Landscapes and their correlation functions. *J. Math. Chem.*, 20:1–45, 1996.

[158] P. F. Stadler. Spectral landscape theory. In J. P. Crutchfield and P. Schuster, editors, *Evolutionary Dynamics—Exploring the Interplay*

*of Selection, Neutrality, Accident, and Function*, pages 231–272. Oxford University Press, 2002.

[159] P. F. Stadler. Canonical approximation of landscapes. Technical Report SFI 94-09-51, Santa Fe Institute, Santa Fe, NM, USA, 1994.

[160] P. F. Stadler and R. Happel. Random field models for fitness landscapes. *J. Math. Biol.*, 38:435–478, 1999.

[161] P. F. Stadler and R. Happel. Correlation structure of the landscape of the graph-bipartitioning-problem. *J. Phys. A: Math. Gen.*, 25:3103–3110, 1992.

[162] P. F. Stadler and B. Krakhofer. Local minima of $p$-spin models. *Rev. Mex. Fis.*, 42:355–363, 1996.

[163] P. F. Stadler and W. Schnabl. The landscape of the travelling salesman problem. *Phys. Lett. A*, 161:337–344, 1992.

[164] P. F. Stadler, R. Seitz, and G. P. Wagner. Evolvability of complex characters: Population dependent Fourier decomposition of fitness landscapes over recombination spaces. *Bull. Math. Biol.*, 62:399–428, 2000.

[165] M. W. Strickberger. *Evolution*. Jones and Bartlett, Boston MA, 1990.

[166] F. Tanaka and S. F. Edwards. Analytic theory of the ground state properties of a spin glass: I. Ising spin glass. *J. Phys. F: Metal Phys.*, 10:2769–2778, 1980.

[167] D. J. Thouless, P. W. Anderson, and R. G. Palmer. Solution of 'Solvable model of a spin glass'. *Phil. Mag.*, 35:593–601, 1977.

[168] N. Trinajstić. *Chemical Graph Theory*. CRC Press, Boca Raton, FL, 2nd edition, 1992.

[169] H. van der Holst. *Topological and Spectral Graph Characterizations*. PhD thesis, Universiteit van Amsterdam, 1996.

[170] H. van der Holst, L. Lovász, and S. A. The Colin de Verdière graph parameter. In *Graph Theory and Computational Biology (Balatonlelle, 1996)*, Mathematical Sudies, pages 29–85. Janos Bolyai Math. Soc., Budapest, 1999.

[171] J. L. Walsh. A closed set of normal orthogonal functions. *Amer. J. Math.*, 45:5–24, 1923.

[172] J. Wang and M. Xu. Quasi-Abelian Cayley graphs and Parsons graphs. *Eur. J. Comb.*, 18:597–600, 1997.

[173] E. D. Weinberger. Correlated and uncorrelated fitness landscapes and how to tell the difference. *Biol. Cybern.*, 63:325–336, 1990.

[174] E. D. Weinberger. Local properties of Kauffman's N-k model: A tunably rugged energy landscape. *Phys. Rev. A*, 44:6399–6413, 1991.

[175] D. B. West. *Introduction to Graph Theory*. Prentice Hall, New York, 1996.

[176] H. Whitney. Congruent graphs and the connectivity of graphs. *Amer. J. Math.*, 54:150–168, 1932.

[177] S. Wright. The roles of mutation, inbreeding, crossbreeeding and selection in evolution. In D. F. Jones, editor, *Proceedings of the Sixth*

*International Congress on Genetics*, volume 1, pages 356–366, New York, 1932. Brooklyn Botanic Gardens.

[178] S. Wright. "Surfaces" of selective value. *Proc. Nat. Acad. Sci. USA*, 58: 165–172, 1967.

[179] B. Zgrablic. On adjacency-transitive graphs. *Discrete Math.*, 182:321–332, 1998.

[180] G. M. Ziegler. *Lectures on Polytopes*, volume 152 of *Graduate Texts in Mathematics*. Springer-Verlag, New York, 1995.

# Index

# Lecture Notes in Mathematics

For information about earlier volumes
please contact your bookseller or Springer
LNM Online archive: springerlink.com

Vol. 1825: J. H. Bramble, A. Cohen, W. Dahmen, Multiscale Problems and Methods in Numerical Simulations, Martina Franca, Italy 2001. Editor: C. Canuto (2003)

Vol. 1826: K. Dohmen, Improved Bonferroni Inequalities via Abstract Tubes. Inequalities and Identities of Inclusion-Exclusion Type. VIII, 113 p, 2003.

Vol. 1827: K. M. Pilgrim, Combinations of Complex Dynamical Systems. IX, 118 p, 2003.

Vol. 1828: D. J. Green, Gröbner Bases and the Computation of Group Cohomology. XII, 138 p, 2003.

Vol. 1829: E. Altman, B. Gaujal, A. Hordijk, Discrete-Event Control of Stochastic Networks: Multimodularity and Regularity. XIV, 313 p, 2003.

Vol. 1830: M. I. Gil', Operator Functions and Localization of Spectra. XIV, 256 p, 2003.

Vol. 1831: A. Connes, J. Cuntz, E. Guentner, N. Higson, J. E. Kaminker, Noncommutative Geometry, Martina Franca, Italy 2002. Editors: S. Doplicher, L. Longo (2004)

Vol. 1832: J. Azéma, M. Émery, M. Ledoux, M. Yor (Eds.), Séminaire de Probabilités XXXVII (2003)

Vol. 1833: D.-Q. Jiang, M. Qian, M.-P. Qian, Mathematical Theory of Nonequilibrium Steady States. On the Frontier of Probability and Dynamical Systems. IX, 280 p, 2004.

Vol. 1834: Yo. Yomdin, G. Comte, Tame Geometry with Application in Smooth Analysis. VIII, 186 p, 2004.

Vol. 1835: O.T. Izhboldin, B. Kahn, N.A. Karpenko, A. Vishik, Geometric Methods in the Algebraic Theory of Quadratic Forms. Summer School, Lens, 2000. Editor: J.-P. Tignol (2004)

Vol. 1836: C. Năstăsescu, F. Van Oystaeyen, Methods of Graded Rings. XIII, 304 p, 2004.

Vol. 1837: S. Tavaré, O. Zeitouni, Lectures on Probability Theory and Statistics. Ecole d'Eté de Probabilités de Saint-Flour XXXI-2001. Editor: J. Picard (2004)

Vol. 1838: A.J. Ganesh, N.W. O'Connell, D.J. Wischik, Big Queues. XII, 254 p, 2004.

Vol. 1839: R. Gohm, Noncommutative Stationary Processes. VIII, 170 p, 2004.

Vol. 1840: B. Tsirelson, W. Werner, Lectures on Probability Theory and Statistics. Ecole d'Eté de Probabilités de Saint-Flour XXXII-2002. Editor: J. Picard (2004)

Vol. 1841: W. Reichel, Uniqueness Theorems for Variational Problems by the Method of Transformation Groups (2004)

Vol. 1842: T. Johnsen, A. L. Knutsen, K₃ Projective Models in Scrolls (2004)

Vol. 1843: B. Jefferies, Spectral Properties of Noncommuting Operators (2004)

Vol. 1844: K.F. Siburg, The Principle of Least Action in Geometry and Dynamics (2004)

Vol. 1845: Min Ho Lee, Mixed Automorphic Forms, Torus Bundles, and Jacobi Forms (2004)

Vol. 1846: H. Ammari, H. Kang, Reconstruction of Small Inhomogeneities from Boundary Measurements (2004)

Vol. 1847: T.R. Bielecki, T. Björk, M. Jeanblanc, M. Rutkowski, J.A. Scheinkman, W. Xiong, Paris-Princeton Lectures on Mathematical Finance 2003 (2004)

Vol. 1848: M. Abate, J. E. Fornaess, X. Huang, J. P. Rosay, A. Tumanov, Real Methods in Complex and CR Geometry, Martina Franca, Italy 2002. Editors: D. Zaitsev, G. Zampieri (2004)

Vol. 1849: Martin L. Brown, Heegner Modules and Elliptic Curves (2004)

Vol. 1850: V. D. Milman, G. Schechtman (Eds.), Geometric Aspects of Functional Analysis. Israel Seminar 2002-2003 (2004)

Vol. 1851: O. Catoni, Statistical Learning Theory and Stochastic Optimization (2004)

Vol. 1852: A.S. Kechris, B.D. Miller, Topics in Orbit Equivalence (2004)

Vol. 1853: Ch. Favre, M. Jonsson, The Valuative Tree (2004)

Vol. 1854: O. Saeki, Topology of Singular Fibers of Differential Maps (2004)

Vol. 1855: G. Da Prato, P.C. Kunstmann, I. Lasiecka, A. Lunardi, R. Schnaubelt, L. Weis, Functional Analytic Methods for Evolution Equations. Editors: M. Iannelli, R. Nagel, S. Piazzera (2004)

Vol. 1856: K. Back, T.R. Bielecki, C. Hipp, S. Peng, W. Schachermayer, Stochastic Methods in Finance, Bressanone/Brixen, Italy, 2003. Editors: M. Fritelli, W. Runggaldier (2004)

Vol. 1857: M. Émery, M. Ledoux, M. Yor (Eds.), Séminaire de Probabilités XXXVIII (2005)

Vol. 1858: A.S. Cherny, H.-J. Engelbert, Singular Stochastic Differential Equations (2005)

Vol. 1859: E. Letellier, Fourier Transforms of Invariant Functions on Finite Reductive Lie Algebras (2005)

Vol. 1860: A. Borisyuk, G.B. Ermentrout, A. Friedman, D. Terman, Tutorials in Mathematical Biosciences I. Mathematical Neurosciences (2005)

Vol. 1861: G. Benettin, J. Henrard, S. Kuksin, Hamiltonian Dynamics – Theory and Applications, Cetraro, Italy, 1999. Editor: A. Giorgilli (2005)

Vol. 1862: B. Helffer, F. Nier, Hypoelliptic Estimates and Spectral Theory for Fokker-Planck Operators and Witten Laplacians (2005)

Vol. 1863: H. Führ, Abstract Harmonic Analysis of Continuous Wavelet Transforms (2005)

Vol. 1864: K. Efstathiou, Metamorphoses of Hamiltonian Systems with Symmetries (2005)

Vol. 1865: D. Applebaum, B.V. R. Bhat, J. Kustermans, J. M. Lindsay, Quantum Independent Increment Processes I. From Classical Probability to Quantum Stochastic Calculus. Editors: M. Schürmann, U. Franz (2005)

Vol. 1866: O.E. Barndorff-Nielsen, U. Franz, R. Gohm, B. Kümmerer, S. Thorbjønsen, Quantum Independent Increment Processes II. Structure of Quantum Lévy Processes, Classical Probability, and Physics. Editors: M. Schürmann, U. Franz, (2005)

Vol. 1867: J. Sneyd (Ed.), Tutorials in Mathematical Biosciences II. Mathematical Modeling of Calcium Dynamics and Signal Transduction. (2005)

Vol. 1868: J. Jorgenson, S. Lang, Posₙ(R) and Eisenstein Series. (2005)

Vol. 1869: A. Dembo, T. Funaki, Lectures on Probability Theory and Statistics. Ecole d'Eté de Probabilités de Saint-Flour XXXIII-2003. Editor: J. Picard (2005)

Vol. 1870: V.I. Gurariy, W. Lusky, Geometry of Müntz Spaces and Related Questions. (2005)

Vol. 1871: P. Constantin, G. Gallavotti, A.V. Kazhikhov, Y. Meyer, S. Ukai, Mathematical Foundation of Turbulent Viscous Flows, Martina Franca, Italy, 2003. Editors: M. Cannone, T. Miyakawa (2006)

Vol. 1872: A. Friedman (Ed.), Tutorials in Mathematical Biosciences III. Cell Cycle, Proliferation, and Cancer (2006)

Vol. 1873: R. Mansuy, M. Yor, Random Times and Enlargements of Filtrations in a Brownian Setting (2006)

Vol. 1874: M. Yor, M. Émery (Eds.), In Memoriam Paul-André Meyer - Séminaire de Probabilités XXXIX (2006)

Vol. 1875: J. Pitman, Combinatorial Stochastic Processes. Ecole d'Eté de Probabilités de Saint-Flour XXXII-2002. Editor: J. Picard (2006)

Vol. 1876: H. Herrlich, Axiom of Choice (2006)

Vol. 1877: J. Steuding, Value Distributions of L-Functions (2007)

Vol. 1878: R. Cerf, The Wulff Crystal in Ising and Percolation Models, Ecole d'Eté de Probabilités de Saint-Flour XXXIV-2004. Editor: Jean Picard (2006)

Vol. 1879: G. Slade, The Lace Expansion and its Applications, Ecole d'Eté de Probabilités de Saint-Flour XXXIV-2004. Editor: Jean Picard (2006)

Vol. 1880: S. Attal, A. Joye, C.-A. Pillet, Open Quantum Systems I, The Hamiltonian Approach (2006)

Vol. 1881: S. Attal, A. Joye, C.-A. Pillet, Open Quantum Systems II, The Markovian Approach (2006)

Vol. 1882: S. Attal, A. Joye, C.-A. Pillet, Open Quantum Systems III, Recent Developments (2006)

Vol. 1883: W. Van Assche, F. Marcellàn (Eds.), Orthogonal Polynomials and Special Functions, Computation and Application (2006)

Vol. 1884: N. Hayashi, E.I. Kaikina, P.I. Naumkin, I.A. Shishmarev, Asymptotics for Dissipative Nonlinear Equations (2006)

Vol. 1885: A. Telcs, The Art of Random Walks (2006)

Vol. 1886: S. Takamura, Splitting Deformations of Degenerations of Complex Curves (2006)

Vol. 1887: K. Habermann, L. Habermann, Introduction to Symplectic Dirac Operators (2006)

Vol. 1888: J. van der Hoeven, Transseries and Real Differential Algebra (2006)

Vol. 1889: G. Osipenko, Dynamical Systems, Graphs, and Algorithms (2006)

Vol. 1890: M. Bunge, J. Funk, Singular Coverings of Toposes (2006)

Vol. 1891: J.B. Friedlander, D.R. Heath-Brown, H. Iwaniec, J. Kaczorowski, Analytic Number Theory, Cetraro, Italy, 2002. Editors: A. Perelli, C. Viola (2006)

Vol. 1892: A. Baddeley, I. Bárány, R. Schneider, W. Weil, Stochastic Geometry, Martina Franca, Italy, 2004. Editor: W. Weil (2007)

Vol. 1893: H. Hanßmann, Local and Semi-Local Bifurcations in Hamiltonian Dynamical Systems, Results and Examples (2007)

Vol. 1894: C.W. Groetsch, Stable Approximate Evaluation of Unbounded Operators (2007)

Vol. 1895: L. Molnár, Selected Preserver Problems on Algebraic Structures of Linear Operators and on Function Spaces (2007)

Vol. 1896: P. Massart, Concentration Inequalities and Model Selection, Ecole d'Eté de Probabilités de Saint-Flour XXXIII-2003. Editor: J. Picard (2007)

Vol. 1897: R. Doney, Fluctuation Theory for Lévy Processes, Ecole d'Eté de Probabilités de Saint-Flour XXXV-2005. Editor: J. Picard (2007)

Vol. 1898: H.R. Beyer, Beyond Partial Differential Equations, On linear and Quasi-Linear Abstract Hyperbolic Evolution Equations (2007)

Vol. 1899: Séminaire de Probabilités XL. Editors: C. Donati-Martin, M. Émery, A. Rouault, C. Stricker (2007)

Vol. 1900: E. Bolthausen, A. Bovier (Eds.), Spin Glasses (2007)

Vol. 1901: O. Wittenberg, Intersections de deux quadriques et pinceaux de courbes de genre 1, Intersections of Two Quadrics and Pencils of Curves of Genus 1 (2007)

Vol. 1902: A. Isaev, Lectures on the Automorphism Groups of Kobayashi-Hyperbolic Manifolds (2007)

Vol. 1903: G. Kresin, V. Maz'ya, Sharp Real-Part Theorems (2007)

Vol. 1904: P. Giesl, Construction of Global Lyapunov Functions Using Radial Basis Functions (2007)

Vol. 1905: C. Prévôt, M. Röckner, A Concise Course on Stochastic Partial Differential Equations (2007)

Vol. 1906: T. Schuster, The Method of Approximate Inverse: Theory and Applications (2007)

Vol. 1907: M. Rasmussen, Attractivity and Bifurcation for Nonautonomous Dynamical Systems (2007)

Vol. 1908: T.J. Lyons, M. Caruana, T. Lévy, Differential Equations Driven by Rough Paths, Ecole d'Eté de Probabilités de Saint-Flour XXXIV-2004. (2007)

Vol. 1909: H. Akiyoshi, M. Sakuma, M. Wada, Y. Yamashita, Punctured Torus Groups and 2-Bridge Knot Groups (I) (2007)

Vol. 1910: V.D. Milman, G. Schechtman (Eds.), Geometric Aspects of Functional Analysis. Israel Seminar 2004-2005 (2007)

Vol. 1911: A. Bressan, D. Serre, M. Williams, K. Zumbrun, Hyperbolic Systems of Balance Laws. Lectures given at the C.I.M.E. Summer School held in Cetraro, Italy, July 14–21, 2003. Editor: P. Marcati (2007)

Vol. 1912: V. Berinde, Iterative Approximation of Fixed Points (2007)

Vol. 1913: J.E. Marsden, G. Misiołek, J.-P. Ortega, M. Perlmutter, T.S. Ratiu, Hamiltonian Reduction by Stages (2007)

Vol. 1914: G. Kutyniok, Affine Density in Wavelet Analysis (2007)

Vol. 1915: T. Bıyıkoğlu, J. Leydold, P.F. Stadler, Laplacian Eigenvectors of Graphs. Perron-Frobenius and Faber-Krahn Type Theorems (2007)

Vol. 1916: C. Villani, F. Rezakhanlou, Entropy Methods for the Boltzmann Equation. Editors: F. Golse, S. Olla (forthcoming)

Vol. 1917: I. Veselić, Existence and Regularity Properties of the Integrated Density of States of Random Schrödinger (2007)

Vol. 1918: B. Roberts, R. Schmidt, Local Newforms for GSp(4) (2007)

Vol. 1919: R.A. Carmona, I. Ekeland, A. Kohatsu-Higa, J.-M. Lasry, P.-L. Lions, H. Pham, E. Taflin, Paris-Princeton Lectures on Mathematical Finance 2004. Editors: R.A. Carmona, E. Çinlar, I. Ekeland, E. Jouini, J.A. Scheinkman, N. Touzi (2007)

## Recent Reprints and New Editions

Vol. 1618: G. Pisier, Similarity Problems and Completely Bounded Maps. 1995 – 2nd exp. edition (2001)

Vol. 1629: J.D. Moore, Lectures on Seiberg-Witten Invariants. 1997 – 2nd edition (2001)

Vol. 1638: P. Vanhaecke, Integrable Systems in the realm of Algebraic Geometry. 1996 – 2nd edition (2001)

Vol. 1702: J. Ma, J. Yong, Forward-Backward Stochastic Differential Equations and their Applications. 1999 – Corr. 3rd printing (2007)

Vol. 830: J.A. Green, Polynomial Representations of $GL_n$, with an Appendix on Schensted Correspondence and Littelmann Paths by K. Erdmann, J.A. Green and M. Schocker 1980 – 2nd corr. and augmented edition (2007)

Printed in the United States
By Bookmasters